Cambridge Elements ≡

Elements in the Structure and Dynamics of Complex Networks
edited by
Guido Caldarelli
Ca' Foscari University of Venice

DESCRIPTIVE VS. INFERENTIAL COMMUNITY DETECTION IN NETWORKS

Pitfalls, Myths, and Half-Truths

Tiago P. Peixoto
Central European University, Vienna

CAMBRIDGE
UNIVERSITY PRESS

Shaftesbury Road, Cambridge CB2 8EA, United Kingdom

One Liberty Plaza, 20th Floor, New York, NY 10006, USA

477 Williamstown Road, Port Melbourne, VIC 3207, Australia

314–321, 3rd Floor, Plot 3, Splendor Forum, Jasola District Centre,
New Delhi – 110025, India

103 Penang Road, #05–06/07, Visioncrest Commercial, Singapore 238467

Cambridge University Press is part of Cambridge University Press & Assessment,
a department of the University of Cambridge.

We share the University's mission to contribute to society through the pursuit of
education, learning and research at the highest international levels of excellence.

www.cambridge.org
Information on this title: www.cambridge.org/9781009113007
DOI: 10.1017/9781009118897

First published 2023

A catalogue record for this publication is available from the British Library.

ISBN 978-1-009-11300-7 Paperback
ISSN 2516-5763 (online)
ISSN 2516-5755 (print)

Descriptive vs. Inferential Community Detection in Networks

Pitfalls, Myths, and Half-Truths

Elements in the Structure and Dynamics of Complex Networks

DOI: 10.1017/9781009118897
First published online: May 2023

Tiago P. Peixoto
Central European University, Vienna

Author for correspondence: Tiago P. Peixoto, peixotot@ceu.edu

Abstract: Community detection is one of the most important methodological fields of network science, and one which has attracted a significant amount of attention over the past decades. This area deals with the automated division of a network into fundamental building blocks, with the objective of providing a summary of its large-scale structure. Despite the importance and widespread adoption of community detection there is a noticeable gap between what is arguably the state-of-the-art and the methods that are actually used in practice in a variety of fields. This Element attempts to address this discrepancy by dividing existing methods according to whether they have a "descriptive" or an "inferential" goal. While descriptive methods find patterns in networks based on context-dependent notions of community structure, inferential methods articulate a precise generative model, and attempt to fit it to data. In this way, they are able to provide insights into the mechanisms of network formation, and separate structure from randomness in a manner supported by statistical evidence. We review how employing descriptive methods with inferential aims is riddled with pitfalls and misleading answers, and thus should be in general avoided. We argue that inferential methods are more typically aligned with clearer scientific questions, yield more robust results, and should be in many cases preferred. We attempt to dispel some myths and half-truths often believed when community detection is employed in practice, in an effort to improve both the use of such methods as well as the interpretation of their results. This title is also available as Open Access on Cambridge Core.

Keywords: community detection, stochastic block models, network clustering, statistical inference, Bayesian inference

ISBNs: 9781009113007 (PB), 9781009118897 (OC)
ISSNs: 2516-5763 (online), 2516-5755 (print)

Contents

1 Introduction   1

2 Descriptive vs. inferential community detection   2

3 Modularity maximization considered harmful   22

4 Myths, pitfalls, and half-truths   29

5 Conclusion   65

References   66

1 Introduction

Community detection is the task of dividing a network — typically one which is large — into many smaller groups of nodes that have a similar contribution to the overall network structure. With such a division, we can better summarize the large-scale structure of a network by describing how these groups are connected, rather than describing each individual node. This simplified description can be used to digest an otherwise intractable representation of a large system, providing insight into its most important patterns, how these patterns relate to its function, and the underlying mechanisms responsible for its formation.

Because of its important role in network science, community detection has attracted substantial attention from researchers, specially in the last 20 years, culminating in an abundant literature (see Refs. [1, 2] for a review). This field has developed significantly from its early days, specially over the last 10 years, during which the focus has been shifting towards methods that are based on statistical inference (see e.g. Refs. [3–5]).

Despite this shift in the state-of-the-art, there remains a significant gap between the best practices and the adopted practices in the use of community detection for the analysis of network data. It is still the case that some of the earliest methods proposed remain in widespread use, despite their many serious shortcomings that have been uncovered over the years. Most of these problems have been addressed with more recent methods, that also contributed to a much deeper theoretical understanding of the problem of community detection [3, 4, 6, 7].

Nevertheless, some misconceptions remain and are still promoted. Here we address some of the more salient ones, in an effort to dispel them. These misconceptions are not uniformly shared; and those that pay close attention to the literature will likely find few surprises here. However, it is possible that many researchers employing community detection are simply unaware of the issues with the methods being used. Perhaps even more commonly, there are those that are in fact aware of them, but not of their actual solutions, or the fact that some supposed countermeasures are ineffective.

Throughout the following we will avoid providing "black box" recipes to be followed uncritically, and instead try as much as possible to frame the issues within a theoretical framework, such that the criticisms and solutions can be justified in a principled manner.

We will set the stage by making a fundamental distinction between "descriptive" and "inferential" community detection approaches. As others have emphasized before [8], community detection can be performed with many goals in mind, and this will dictate which methods are most appropriate. We will provide a simple "litmus test" that can be used to determine which overall approach is more adequate, based on whether our goal is to seek inferential interpretations. We will then move to a more focused critique of the method that is arguably the most widely employed — modularity maximization. This method has an exemplary character, since it contains all possible pitfalls of using descriptive methods for inferential aims. We will then follow with a discussion of myths, pitfalls, and half-truths that obstruct a more effective analysis of community structure in networks.

(We will not give a throughout technical introduction to inferential community detection methods, which can be obtained instead in Ref. [5]. For a practical guide on how to use various inferential methods, readers are referred to the detailed HOWTO[1] available as part of the graph-tool Python library [9].)

2 Descriptive vs. inferential community detection

At a very fundamental level, community detection methods can be divided into two main categories: "descriptive" and "inferential."

Descriptive methods attempt to find communities according to some context-dependent notion of a good division of the network into groups. These notions are based on the patterns that can be identified in the network via an exhaustive algorithm, but without taking into consideration the possible rules that were used to create the patterns uncovered. These patterns are used only to *describe* the network, not to explain it. Usually, these approaches do not articulate precisely what constitutes community structure to begin with, and focus instead only on how to detect such patterns. For this kind of method, concepts of statistical significance, parsimony, and generalizability are usually not evoked.

Inferential methods, on the other hand, start with an explicit definition of what constitutes community structure, via a generative model for the network. This model describes how a *latent* (i.e. not observed) partition of the nodes would affect the placement of the edges. The inference consists on reversing this procedure to determine which node partitions are more likely to have been responsible for the observed network. The result of this is a "fit" of a model to data, that can be used as a tentative explanation of how the network came to be. The concepts of statistical significance, parsimony, and generalizability arise naturally and can be quantitatively assessed in this context.

---

[1] Available at https://graph-tool.skewed.de/static/doc/demos/inference/inference.html.

Descriptive community detection methods are by far the most numerous, and those that are in most widespread use. However, this contrasts with the current state-of-the-art, which is composed in large part of inferential approaches. Here we point out the major differences between them and discuss how to decide which is more appropriate, and also why one should in general favor the inferential varieties whenever the objective is to derive generative interpretations from data.

2.1 Describing vs. explaining

We begin by observing that descriptive clustering approaches are the methods of choice in certain contexts. For instance, such approaches arise naturally when the objective is to divide a network into two or more parts as a means to solve a variety of optimization problems. Arguably, the most classic example of this is the design of very large scale integrated (VLSI) circuits [10]. The task is to combine from up to billions of transistors into a single physical microprocessor chip. Transistors that connect to each other must be placed together to take less space, consume less power, reduce latency, and reduce the risk of crosstalk with other nearby connections. To achieve this, the initial stage of a VLSI process involves the partitioning of the circuit into many smaller modules with few connections between them, in a manner that enables their efficient spatial placement, i.e. by positioning the transistors in each module close together and those in different modules farther apart.

Another notable example is parallel task scheduling, a problem that appears in computer science and operations research. The objective is to distribute processes (i.e. programs, or tasks in general) between different processors, so they can run at the same time. Since processes depend on the partial results of other processes, this forms a dependency network, which then needs to be divided such that the number of dependencies across processors is minimized. The optimal division is the one where all tasks are able to finish in the shortest time possible.

Both examples above, and others, have motivated a large literature on "graph partitioning" dating back to the 70s [11–13], which covers a family of problems that play an important role in computer science and algorithmic complexity theory.

Although reminiscent of graph partitioning, and sharing with it many algorithmic similarities, community detection is used more broadly with a different goal [1, 2]. Namely, the objective is to perform *data analysis*, where one wants to extract scientific understanding from empirical observations. The communities identified are usually directly used for representation and/or interpretation of the data, rather than as a mere device to solve a particular optimization problem. In this context, a merely descriptive approach will fail at giving us a meaningful insight into the data, and can be misleading, as we will discuss in the following.

We illustrate the difference between descriptive and inferential approaches in Fig. 1. We first make an analogy with the famous "face" seen on images of the *Cydonia Mensae* region of the planet Mars. A merely descriptive account of the image can be made by identifying the facial features seen, which most people immediately recognize. However, an inferential description of the same image would seek instead to *explain* what is being seen. The process of explanation must invariably involve at its core an application of the law of parsimony, or **Occam's razor**. This principle predicates that when considering two hypotheses compatible with an observation, the simplest one must prevail. Employing this logic results in the conclusion that what we are seeing is in fact a regular mountain, without denying that it looks like a face in that picture and instead acknowledging that it does so accidentally. In other words, the "facial" description is not useful as an explanation, as it emerges out of random features rather than exposing any underlying mechanism.

Going out of the analogy and back to the problem of community detection, in Fig. 1(c) and (d) we see a descriptive and an inferential account of an example network, respectively. The descriptive one is a division of the nodes into 13 assortative communities, which would be identified with many descriptive community detection methods available in the literature. Indeed, we can inspect visually that these groups form assortative communities,[2] and most people would agree that these communities are really there, according to most definitions in use: these are groups of nodes with many more internal edges than external ones. However, an inferential account of the same network would reveal something else altogether. Specifically, it would explain this network as the outcome of a process where the edges are placed at random, without the existence of any communities. The communities that we see in Fig. 1(c) are just a byproduct of this random process, and therefore carry no explanatory power. In fact, this is exactly how the network in this example was generated, i.e. by choosing a specific degree sequence and connecting the edges uniformly at random.

In Fig. 2(a) we illustrate in more detail how the network in Fig. 1 was generated: The degrees of the nodes are fixed, forming "stubs" or "half-edges," which are then paired uniformly at random forming the edges of the network.[3]

---

[2] See Sec. 4.6 for possible pitfalls with relying on visual inspections.

[3] This uniform pairing will typically also result in the occurrence of pairs of nodes of degree one connected together in their own connected component. We consider instances of the process

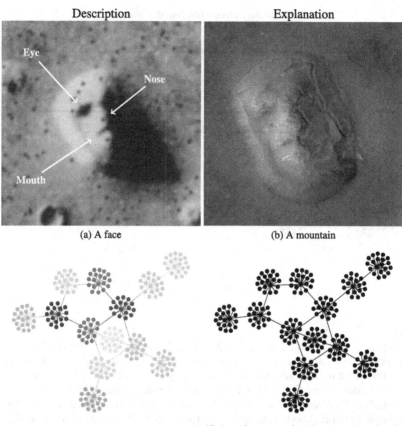

Description   Explanation

(a) A face   (b) A mountain

(c) A network with 13 communities   (d) A random network with a prescribed degree
                                         sequence, and no community structure.

Figure 1 Difference between descriptive and inferential approaches to data analysis. As an analogy, in panels (a) and (b) we see two representations of the *Cydonia Mensae* region on Mars. Panel (a) is a descriptive account of what we see in the picture, namely a face. Panel (b) is an inferential representation of what lies behind it, namely a mountain (this is a more recent image of the same region with a higher resolution to represent an inferential interpretation of the figure in panel (a)). More concretely for the problem of community detection, in panels (c) and (d) we see two representations of the same network. Panel (c) shows a descriptive division into 13 assortative communities. In panel (d) we see an inferential representation as a degree-constrained random network, with no communities, since this is a more likely model of how this network was formed (see Fig. 2).

In Fig. 2(b), like in Fig. 1, the node colors show the partition found with descriptive community detection methods. However, this network division carries no

---

where this does not happen for visual clarity in Fig. 2(c) and (d), but without sacrificing its main message.

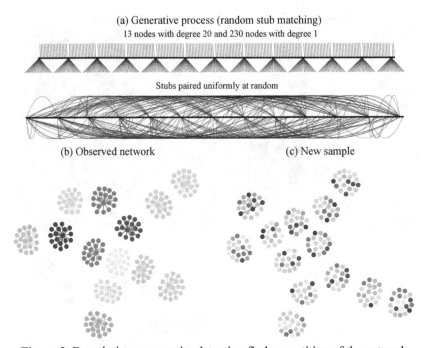

(a) Generative process (random stub matching)

13 nodes with degree 20 and 230 nodes with degree 1

Stubs paired uniformly at random

(b) Observed network   (c) New sample

Figure 2 Descriptive community detection finds a partition of the network according to an arbitrary criterion that bears in general no relation to the rules that were used to generate it. In (a) is shown the generative model we consider, where first a degree sequence is given to the nodes (forming "stubs", or "half-edges") which then are paired uniformly at random, forming a graph. In (b) is shown a realization of this model. The node colors show the partition found with virtually any descriptive community detection method. In (c) is shown another network sampled from the same model, together with the same partition found in (b), which is completely uncorrelated with the new apparent communities seen, since they are the mere byproduct of the random placement of the edges. An inferential approach would find only a single community in both (b) and (c), since no partition of the nodes is relevant for the underlying generative model.

explanatory power beyond what is contained in the degree sequence of the network, since it is generated otherwise uniformly at random. This becomes evident in Fig. 2(c), where we show another network sampled from the same generative process, i.e. another random pairing, but partitioned according to the same division as in Fig. 2(b). Since the nodes are paired uniformly at random, constrained only by their degree, this will create new apparent "communities" that are always uncorrelated with one another. Like the "face" on Mars, they can be seen and described, but they cannot (plausibly) explain how the network came to be.

We emphasize that the communities found in Fig. 2(b) are indeed really there from a descriptive point of view, and they can in fact be useful for a variety of tasks. For example, the *cut* given by the partition, i.e. the number of edges that go between different groups, is only 13, which means that we need only to remove this number of edges to break the network into (in this case) 13 smaller components. Depending on context, this kind of information can be used to prevent a widespread epidemic, hinder undesired communication, or, as we have already discussed, distribute tasks among processors and design a microchip. However, what these communities *cannot* be used for is to *explain* the data. In particular, a conclusion that would be completely incorrect is that the nodes that belong to the same group would have a larger probability of being connected between themselves. As shown in Fig. 2(a), this is clearly not the case, as the observed "communities" arise by pure chance, without any preference between the nodes.

2.2 To infer or to describe? A litmus test

Given the above differences, and the fact that both inferential and descriptive approaches have their uses depending on context, we are left with the question: Which approach is more appropriate for a given task at hand? In order to help answering this question, for any given context, it is useful to consider the following "litmus test":

> **— *Litmus test: to infer or to describe?* —**
>
> Q: "Would the usefulness of our conclusions change if we learn, after obtaining the communities, that the network being analyzed is maximally random?"
>
> If the answer is "yes," then an inferential approach is needed.
>
> If the answer is "no," then an inferential approach is not required.

If the answer to the above question is "yes," then an inferential approach is warranted, since the conclusions depend on an interpretation of how the data were generated. Otherwise, a purely descriptive approach may be appropriate since considerations about generative processes are not relevant.

It is important to understand that the relevant question in this context is not whether the network being analyzed is *actually* maximally random,[4] since this

---

[4] "Maximally random" here means that, conditioned on some global or local constraints, like the number of edges or the node degrees, the placement of the edges is done in uniformly at random. In other words, the network is sampled from a maximum-entropy model constrained in a manner unrelated to community structure, such that whatever communities we may ascribe to the nodes could have played no role in the placement of the edges.

is rarely the case for empirical networks. Instead, considering this hypothetical scenario serves as a test to evaluate if our task requires us to separate between actual latent community structures (i.e. those that are responsible for the network formation), from those that arise completely out of random fluctuations, and hence carry no explanatory power. Furthermore, most empirical networks, even if not maximally random, like most interesting data, are better explained by a mixture of structure and randomness, and a method that cannot tell those apart cannot be used for inferential purposes.

Returning to the VLSI and task scheduling examples we considered in the previous section, it is clear that the answer to the litmus test above would be "no," since it hardly matters how the network was generated and how we should interpret the partition found, as long as the integrated circuit can be manufactured and function efficiently, or the tasks finish in the minimal time. Interpretation and explanations are simply not the primary goals in these cases.[5]

However, it is safe to say that in network data analyses very often the answer to the question above question would be "yes." Typically, community detection methods are used to try to understand the overall large-scale network structure, determine the prevalent mixing patterns, make simplifications and generalizations, all in a manner that relies on statements about what lies behind the data, e.g. whether nodes were more or less likely to be connected to begin with. A majority of conclusions reached would be severely undermined if one would discover that the underlying network is in fact completely random. This means that these analyses suffer the substantial risk of yielding misleading answers when using purely descriptive methods, since they are likely to be *overfitting* the data — i.e. confusing randomness with underlying generative structure.[6]

2.3 Inferring, explaining, and compressing

Inferential approaches to community detection (see Ref. [5] for a detailed introduction) are designed to provide explanations for network data in a principled manner. They are based on the formulation of generative models that include

---

[5] Although this is certainly true at a first instance, we can also argue that properly understanding *why* a certain partition was possible in the first place would be useful for reproducibility and to aid the design of future instances of the problem. For these purposes, an inferential approach would be more appropriate.

[6] We emphasize that the concept of overfitting is intrinsically tied with an inferential goal, i.e. one that involves interpretations about an underlying distribution of probability relating to the network structure. The partitioning of a graph with the objective of producing an efficient chip design cannot overfit, because it remove does not elicit an inferential interpretation. Therefore, whenever we mention that a method overfits, we refer only to the situation where it is being employed with an inferential goal, and that it incorporates a level of detail that cannot be justified by the statistical evidence available in the data.

the notion of community structure in the rules of how the edges are placed. More formally, they are based on the definition of a likelihood $P(A|b)$ for the network A conditioned on a partition b, which describes how the network could have been generated, and the inference is obtained via the posterior distribution, according to Bayes' rule, i.e.

$$P(b|A) = \frac{P(A|b)P(b)}{P(A)},$$ (1)

where $P(b)$ is the prior probability for a partition b. The inference procedure consists in sampling from or maximizing this distribution, which yields the most likely division(s) of the network into groups, according to the statistical evidence available in the data (see Fig. 3).

Overwhelmingly, the models used to infer communities are variations of the stochastic block model (SBM) [14], where in addition to the node partition, it takes the probability of edges being placed between the different groups as an additional set of parameters. A particularly expressive variation is the degree-corrected SBM (DC-SBM) [15], with a marginal likelihood given by [16]

$$P(A|b) = \sum_{e,k} P(A|k,e,b)P(k|e,b)P(e|b),$$ (2)

where $e = \{e_{rs}\}$ is a matrix with elements e_{rs} specifying how many edges go between groups r and s, and $k = \{k_i\}$ are the degrees of the nodes. Therefore, this model specifies that, conditioned on a partition b, first the edge counts e are sampled from a prior distribution $P(e|b)$, followed by the degrees from the prior $P(k|e,b)$, and finally the network is wired together according to the probability $P(A|k,e,b)$, which respects the constraints given by k, e, and b. See Fig. 3(a) for a illustration of this process.

This model formulation includes maximally random networks as special cases — indeed the model we considered in Fig. 2 corresponds exactly to the DC-SBM with a single group. Together with the Bayesian approach, the use of this model will inherently favor a more parsimonious account of the data, whenever it does not warrant a more complex description — amounting to a formal implementation of Occam's razor. This is best seen by making a formal connection with information theory, and noticing that we can write the numerator of Eq. 1 as

$$P(A|b)P(b) = 2^{-\Sigma(A,b)},$$ (3)

where the quantity $\Sigma(A,b)$ is known as the *description length* [17–19] of the network. It is computed as[7]

---

[7] Note that the sum in Eq. 2 vanishes because only one term is non-zero given a fixed network A.

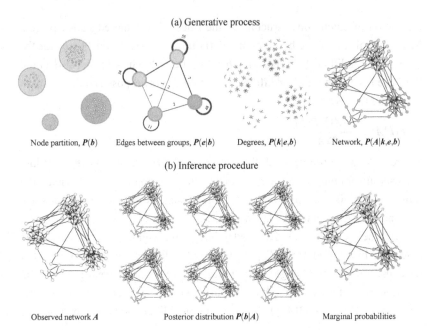

(a) Generative process

Node partition, $P(\boldsymbol{b})$ Edges between groups, $P(\boldsymbol{e}|\boldsymbol{b})$ Degrees, $P(\boldsymbol{k}|\boldsymbol{e},\boldsymbol{b})$ Network, $P(\boldsymbol{A}|\boldsymbol{k},\boldsymbol{e},\boldsymbol{b})$

(b) Inference procedure

Observed network A Posterior distribution $P(\boldsymbol{b}|\boldsymbol{A})$ Marginal probabilities

Figure 3 Inferential community detection considers a generative process (a), where the unobserved model parameters are sampled from prior distributions. In the case of the DC-SBM, these are the priors for the partition $P(\boldsymbol{b})$, the number of edges between groups $P(\boldsymbol{e}|\boldsymbol{b})$, and the node degrees, $P(\boldsymbol{k}|\boldsymbol{e},\boldsymbol{b})$. Finally, the network itself is sampled from its model, $P(\boldsymbol{A}|\boldsymbol{k},\boldsymbol{e},\boldsymbol{b})$. The inference procedure (b) consists on inverting the generative process given an observed network $\boldsymbol{A}$, corresponding to a posterior distribution $P(\boldsymbol{b}|\boldsymbol{A})$, which then can be summarized by a marginal probability that a node belongs to a given group (represented as pie charts on the nodes).

$$\Sigma(\boldsymbol{A},\boldsymbol{b}) = \underbrace{- \log_2 P(\boldsymbol{A}|\boldsymbol{k},\boldsymbol{e},\boldsymbol{b})}_{\mathcal{D}(A|k,e,b)} \underbrace{- \log_2 P(\boldsymbol{k}|\boldsymbol{e},\boldsymbol{b}) - \log_2 P(\boldsymbol{e}|\boldsymbol{b}) - \log_2 P(\boldsymbol{b})}_{\mathcal{M}(k,e,b)}. \quad (4)$$

The second set of terms $\mathcal{M}(\boldsymbol{k},\boldsymbol{e},\boldsymbol{b})$ in the above equation quantifies the amount of information in bits necessary to encode the parameters of the model.[8] The first term $\mathcal{D}(\boldsymbol{A}|\boldsymbol{k},\boldsymbol{e},\boldsymbol{b})$ determines how many bits are necessary to encode the network itself, once the model parameters are known. This means that if Bob wants to communicate to Alice the structure of a network $\boldsymbol{A}$, he first needs to

---

[8] If a value x occurs with probability $P(x)$, this means that in order to transmit it in a communication channel we need to answer at least $-\log_2 P(x)$ yes-or-no questions to decode its value exactly. Therefore we need to answer one yes-or-no question for a value with $P(x) = 1/2$, zero questions for $P(x) = 1$, and $\log_2 N$ questions for uniformly distributed values with $P(x) = 1/N$. This value is called "information content," and essentially measures the degree of "surprise" when encountering a value sampled from a distribution. See Ref. [20] for a thorough but accessible introduction to information theory and its relation to inference.

transmit $\mathscr{M}(k, e, b)$ bits of information to describe the parameters b, e, and k, and then finally transmit the remaining $\mathscr{D}(A|k, e, b)$ bits to describe the network itself. Then, Alice will be able to understand the message by first decoding the parameters (k, e, b) from the first part of the message, and using that knowledge to obtain the network A from the second part, without any errors.

What the above connection shows is that there is a formal equivalence between *inferring* the communities of a network and *compressing* it. This happens because finding the most likely partition b from the posterior $P(b|A)$ is equivalent to minimizing the description length $\Sigma(A, b)$ used by Bob to transmit a message to Alice containing the whole network.

Data compression amounts to a formal implementation of Occam's razor because it penalizes models that are too complicated: if we want to describe a network using many communities, then the model part of the description length $\mathscr{M}(k, e, b)$ will be large, and Bob will need many bits to transmit the model parameters to Alice. However, increasing the complexity of the model will also *reduce* the first term $\mathscr{D}(A|k, e, b)$, since there are fewer networks that are compatible with the bigger set of constraints, and hence the second part of Bob's message will need to be shorter to convey the network itself once the parameters are known. Compression (and hence inference), therefore, is a balancing act between model complexity and quality of fit, where an increase in the former is *only* justified when it results in *an even larger* increase of the second, such that the total description length is minimized.

The reason why the compression approach avoids overfitting the data is due to a powerful fact from information theory, known as Shannon's source coding theorem [21], which states that it is impossible to compress data sampled from a distribution $P(x)$ using fewer bits per symbol than the entropy of the distribution, $H = -\sum_x P(x) \log_2 P(x)$ — indeed, it's a remarkable fact from Shannon's theory that a statement about a single sample (how many bits we need to describe it) is intrinsically connected to the distribution from which it came. Therefore, as the dataset becomes large, it also becomes impossible to compress the data more than can be achieved by using a code that is optimal according to its true distribution. In our context, this means that it is impossible, for example, to compress a maximally random network using a SBM with more than one group.[9] This means, for example, that when encountering an example like in Fig. 2, inferential methods will detect a single community comprising all nodes in the network, since any further division does not provide any increased compression, or equivalently, no augmented explanatory

---

[9] More accurately, this becomes impossible only when the network becomes asymptotically infinite; for finite networks the probability of compression is only vanishingly small.

(a) Observed network　　　　　　　(b) New sample

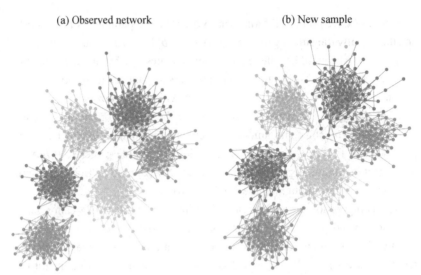

Figure 4 Inferential community detection aims to find a partition of the network according to a fit of a generative model that can explain its structure. In (a) is shown a network sampled from a stochastic block model (SBM) with 6 groups, and where the group assignments were hidden from view. The node colors show the groups found via Bayesian inference of the SBM. In (b) is shown another network sampled from same SBM, together with the same partition found in (a), showing that it carries a substantial explanatory power — very differently from the example in Fig. 2 (c).

power. From the inferential point of view, a partition like Fig. 2(b) *overfits* the data, since it incorporates irrelevant random features — a.k.a. "noise" — into its description.

In Fig. 4(a) is shown an example of the results obtained with an inferential community detection algorithm, for a network sampled from the SBM. As shown in Fig. 4(b), the obtained partitions are still valid when carried over to an independent sample of the model, because the algorithm is capable of separating the general underlying pattern from the random fluctuations. As a consequence of this separability, this kind of algorithm does not find communities in maximally random networks, which are composed only of "noise."

The concept of compression is more generally useful than just avoiding overfitting within a class of models. In fact, the description length gives us a model-agnostic objective criterion to compare different hypotheses for the data generating process according to their plausibility. Namely, since Shannon's theorem tells us that the best compression can be achieved asymptotically only with the true model, then if we are able to find a description length for a network using a particular model, regardless of how it is parametrized, this also means that we have automatically found an *upper bound* on the optimal compression achievable. By formulating different generative models and computing their description length, we have not only an objective criterion to compare them against each other, but we also have a way to limit further what can be obtained with any other model. The result is an overall scale on which different models can be compared, as we move closer to the limit of what can be uncovered for a particular network at hand.

As an example, in Fig. 5 we show the description length values with some models obtained for a protein-protein interaction network for the organism *Meleagris gallopavo* (wild turkey) [22]. In particular, we can see that with the DC-SBM/TC (a version of the model with the addition of triadic closure edges [23]) we can achieve a description length that is far smaller than what would be possible with networks sampled from either the Erdős-Rényi, configuration, or planted partition (a SBM with strictly assortative communities [24]) models, meaning that the inferred model is much closer to the true process that actually generated this network than the alternatives. Naturally, the actual process that generated this network is different from the DC-SBM/TC, and it likely involves, for example, mechanisms of node duplication which are not incorporated into this rather simple model [25]. However, to the extent that the true process leaves statistically significant traces in the network structure,[10] computing the description length according to it should provide further compression when compared to the alternatives.[11] Therefore, we can try to extend or reformulate our models to incorporate features that we hypothesize to be more realistic, and then verify if this in fact the case, knowing that whenever we find a more compressive model, it is moving closer to the true model — or at least to what remains detectable from it for the finite data.

The discussion above glosses over some important technical aspects. For example, it is possible for two (or, in fact, many) models to have the same or very similar description length values. In this case, Occam's razor fails as a criterion to select between them, and we need to consider them collectively as equally valid hypotheses. This means, for example, that we would need to

---

[10] Visually inspecting Fig. 5 reveals what seems to be local symmetries in the network structure, presumably due to gene duplication. These patterns are not exploited by the SBM description, and points indeed to a possible path for further compression.

[11] In Sec. 4.8 we discuss further the usefulness of models like the SBM despite the fact we know they are not the true data generating process.

(a)

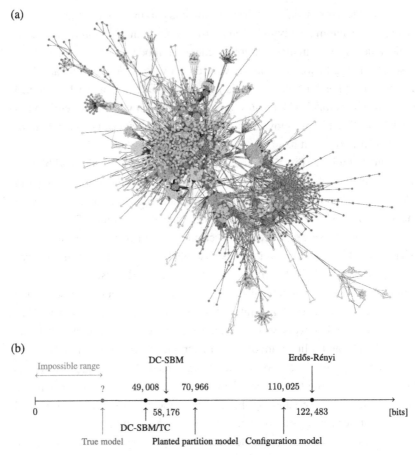

(b)

Impossible range

DC-SBM

Erdős-Rényi

?   49,008   70,966   110,025

0   ↑ 58,176 ↑   122,483   [bits]

DC-SBM/TC

True model   Planted partition model   Configuration model

Figure 5 Compression points towards the true model. (a) Protein-protein interaction network for the organism *Meleagris gallopavo* [22]. The node colors indicate the best partition found with the DC-SBM/TC [23] (there are more groups than colors, so some colors are repeated), and the edge colors indicate whether they are attributed to triadic closure (red) or the DC-SBM (black). (b) Description length values according to different models. The unknown true model must yield a description length value smaller than the DC-SBM/TC, and no other model should be able to provide a superior compression that is statistically significant.

average over them when making specific inferential statements [26] — selecting between them arbitrarily can be interpreted as a form of overfitting. Furthermore, there is obviously no guarantee that the true model can actually be found for any particular data. This is only possible in the asymptotic limit of "sufficient data," which will vary depending on the actual model. Outside of this limit (which is the typical case in empirical settings, in particular when dealing with *sparse* networks [27]), fundamental limits to inference are unavoidable,[12] which means in practice that we will always have limited accuracy and some amount of error in our conclusions. However, when employing compression, these potential errors tend towards overly simple explanations, rather than overly complex ones. Whenever perfect accuracy is not possible, it is difficult to argue in favor of a bias in the opposite direction.

We emphasize that it is not possible to "cheat" when doing compression. For any particular model, the description length will have the same form

$$\Sigma(A, \theta) = \mathscr{D}(A|\theta) + \mathscr{M}(\theta), \tag{5}$$

where θ is some arbitrary set of parameters. If we constrain the model such that it becomes possible to describe the data with a number of bits $\mathscr{D}(A|\theta)$ that is very small, this can only be achieved, in general, by increasing the number of parameters θ, such that the number of bits $\mathscr{M}(\theta)$ required to describe them will also increase. Therefore, there is no generic way to achieve compression that bypasses actually formulating a meaningful hypothesis that matches statistically significant patterns seen in the data. One may wonder, therefore, if there is an automatized way of searching for hypotheses in a manner that guarantees optimal compression. The most fundamental way to formulate this question is to generalize the concept of minimum description length as follows: for any binary string x (representing any measurable data), we define $L(x)$ as the length in bits of the shortest computer program that yields x as an output. The quantity $L(x)$ is known as Kolmogorov complexity [29, 30], and if we would be able to compute it for a binary string representing an observed network, we would be able to determine the "true model" value in Fig. 5, and hence know how far we are from the optimum.[13]

Unfortunately, an important result in information theory is that $L(x)$ is not computable [30]. This means that it is strictly impossible to write a computer program that computes $L(x)$ for any string x.[14] This does not invalidate using

---

[12] A very important result in the context of community detection is the detectability limit of the SBM. As discovered by Decelle et al [6, 28], if a sufficiently large network is sampled from a SBM with a sufficiently weak but nontrivial structure below a specific threshold, it becomes strictly impossible to uncover the true model from this sample.

[13] As mentioned before, this would not necessarily mean that we would be able to find the actual true model in a practical setting with perfect accuracy, since for a finite x there could be many programs of the same minimal length (or close) that generate it.

[14] There are two famous ways to prove this. One is by contradiction: if we assume that we have a program that computes $L(x)$, then we could use it as subroutine to write another program that outputs x with a length smaller than $L(x)$. The other involves undecidabilty: if we enumerate all possible computer programs in order of increasing length and check if their outputs match x, we will eventually find programs that loop indefinitely. Deciding whether a program finishes in finite time is known as the "halting problem," which has been proved to be impossible to

the description length as a criterion to select among alternative models, but it dashes any hope of fully automatizing the discovery of optimal hypotheses.

2.4 Role of inferential approaches in community detection

Inferential approaches based on the SBM have an old history, and were introduced for the study of social networks in the early 80's [14]. But despite such an old age, and having appeared repeatedly in the literature over the years [31–39] (also under different names in other contexts e.g. [40, 41]), they entered the mainstream community detection literature rather late, arguably after the influential paper by Karrer and Newman that introduced the DC-SBM [15] in 2011, at a point where descriptive approaches were already dominating. However, despite the dominance of descriptive methods, the existence of inferential *criteria* was already long noticeable. In fact, in a well-known attempt to systematically compare the quality of a variety of descriptive community detection methods, the authors of Ref. [42] proposed the now so-called Lancichinetti–Fortunato–Radicchi (LFR) benchmark, offered as a more realistic alternative to the simpler Newman-Girvan benchmark [43] introduced earlier. Both are in fact generative models, essentially particular cases of the DC-SBM, containing a "ground truth" community label assignment, against which the results of various algorithms are supposed to be compared. Clearly, this is an inferential evaluation criterion, although, historically, virtually all of the methods compared against that benchmark are descriptive in nature [44] (these studies were conducted mostly before inferential approaches had gained more traction). The use of such a criterion already betrays that the answer to the litmus test considered previously would be "yes," and therefore descriptive approaches are fundamentally unsuitable for the task. In contrast, methods based on statistical inference are not only more principled, but in fact provably optimal in the inferential scenario: an estimation based on the posterior distribution obtained from the true generative model is called "Bayes optimal," since there is no procedure that can, on average, produce results with higher accuracy. The use of this inferential formalism has led to the development of asymptotically optimal algorithms and the identification of sharp transitions in the detectability of planted community structure [6, 45].

The conflation one often finds between descriptive and inferential goals in the literature of community detection likely stems from the fact that while it

---

solve. In general, it cannot be determined if a program reaches an infinite loop in a manner that avoids actually running the program and waiting for it to finish. Therefore, this rather intuitive algorithm to determine $L(x)$ will not necessarily finish for any given string x. For more details see e.g. Refs [29, 30]

is easy to define benchmarks in the inferential setting, it is substantially more difficult to do so in a descriptive setting. Given any descriptive method (modularity maximization [46], Infomap [47], Markov stability [48], etc.) it is usually problematic to determine for which network these methods are optimal (or even if one exists), and what would be a canonical output that would be unambiguously correct. In fact, the difficulty with establishing these fundamental references already serve as evidence that the task itself is ill-defined. On the other hand, taking an inferential route forces one to *start with the right answer*, via a well-specified generative model that articulates what *the communities actually mean* with respect to the network structure. Based on this precise definition, one then *derives* the optimal detection method by employing Bayes' rule.

It is also useful to observe that inferential analyses of aspects of the network other than directly its structure might still be only descriptive of the structure itself. A good example of this is the modelling of dynamics that take place on a network, such as a random walk. This is precisely the case of the Infomap method [47], which models a simulated random walk on a network in an inferential manner, using for that a division of the network into groups. While this approach can be considered inferential with respect to an artificial dynamics, it is still only descriptive when it comes to the actual network structure (and will suffer the same problems, such a finding communities in maximally random networks). Communities found in this way could be useful for particular tasks, such as to identify groups of nodes that would be similarly affected by a diffusion process. This could be used, for example, to prevent or facilitate the diffusion by removing or adding edges between the identified groups. In this setting, the answer to the litmus test above would also be "no," since what is important is how the network "is" (i.e. how a random walk behaves on it), not how it came to be, or if its features are there by chance alone. Once more, the important issue to remember is that the groups identified in this manner cannot be interpreted as having any explanatory power about the network structure itself, and cannot be used reliably to extract inferential conclusions about it. We are firmly in a descriptive, not inferential setting with respect to the network structure.

Another important difference between inferential and descriptive approaches is worth mentioning. Descriptive approaches are often tied to very particular contexts, and cannot be directly compared to one another. This has caused great consternation in the literature, since there is a vast number of such methods, and little robust methodology on how to compare them. Indeed, why should we expect that the modules found by optimizing task scheduling should be comparable to those that optimize the description of a dynamics? In contrast, inferential approaches all share the same underlying context: they attempt to explain the network structure; they vary only in how this is done. They are, therefore, amenable to principled *model selection* procedures [20, 49, 50], designed to evaluate which is the most appropriate fit for any particular network, even if the models used operate with very different parametrizations, as we discussed already in Sec. 2.3. In this situation, the multiplicity of different models available becomes a boon rather than a hindrance, since they all contribute to a bigger toolbox we have at our disposal when trying to understand empirical observations.

Finally, inferential approaches offer additional advantages that make them more suitable as part of a scientific pipeline. In particular, they can be naturally extended to accommodate measurement uncertainties [51–53] — an unavoidable property of empirical data, which descriptive methods almost universally fail to consider. This information can be used not only to propagate the uncertainties to the community assignments [26] but also to reconstruct the missing or noisy measurements of the network itself [37, 54]. Furthermore, inferential approaches can be coupled with even more indirect observations such as time-series on the nodes [55], instead of a direct measurement of the edges of the network, such that the network itself is reconstructed, not only the community structure [56]. All these extensions are possible because inferential approaches give us more than just a division of the network into groups; they give us a model estimate of the network, containing insights about its formation mechanism.

2.5 Behind every description there is an implicit generative model

Descriptive methods of community detection — such as graph partitioning for VLSI [11] or Infomap [47] — are not designed to produce inferential statements about the network structure. They do not need to explicitly articulate a generative model, and the quality of their results should be judged solely against their manifestly noninferential goals, e.g. whether a chip design can be efficiently manufactured in the case of graph partitioning.

Nevertheless, descriptive methods are often employed with inferential aims in practice. This happens, for example, when modularity maximization is used to discover homophilic patterns in a social network, or when Infomap is used to uncover latent communities generated by the LFR benchmark. In these situations, it is useful to consider to what extent can we expect any of these methods reveal meaningful inferential results, despite their intended use.

From a purely mathematical perspective, there is actually no formal distinction between descriptive and inferential methods, because every descriptive method can be mapped to an inferential one, according to some implicit model. Therefore, whenever we are attempting to interpret the results of a descriptive community detection method in an inferential way — i.e. make a statement about how the network came to be — we cannot in fact avoid making *implicit* assumptions about the data generating process that lies behind it. (At first this statement seems to undermine the distinction we have been making between descriptive and inferential methods, but in fact this is not the case, as we will see below.)

It is not difficult to demonstrate that it is possible to formulate any conceivable community detection method as a particular inferential method. Let us consider an arbitrary quality function

$$W(\boldsymbol{A}, \boldsymbol{b}) \in \mathbb{R} \tag{6}$$

which is used to perform community detection via the optimization

$$\boldsymbol{b}^* = \underset{\boldsymbol{b}}{\text{argmax}} \; W(\boldsymbol{A}, \boldsymbol{b}). \tag{7}$$

We can then interpret the quality function $W(\boldsymbol{A}, \boldsymbol{b})$ as the "Hamiltonian" of a posterior distribution

$$P(\boldsymbol{b}|\boldsymbol{A}) = \frac{e^{\beta W(A,b)}}{Z(\boldsymbol{A})}, \tag{8}$$

with normalization $Z(\boldsymbol{A}) = \sum_b e^{\beta W(A,b)}$. By making $\beta \to \infty$ we recover the optimization of Eq. 7, or we may simply try to find the most likely partition according to the posterior, in which case $\beta > 0$ remains an arbitrary parameter. Therefore, employing Bayes' rule in the opposite direction, we obtain the following effective generative model:

$$P(\boldsymbol{A}|\boldsymbol{b}) = \frac{P(\boldsymbol{b}|\boldsymbol{A})P(\boldsymbol{A})}{P(\boldsymbol{b})}, \tag{9}$$

$$= \frac{e^{\beta W(A,b)}}{Z(\boldsymbol{A})} \frac{P(\boldsymbol{A})}{P(\boldsymbol{b})}, \tag{10}$$

where $P(\boldsymbol{A}) = \sum_b P(\boldsymbol{A}|\boldsymbol{b})P(\boldsymbol{b})$ is the marginal distribution over networks, and $P(\boldsymbol{b})$ is the prior distribution for the partition. Due to the normalization of $P(\boldsymbol{A}|\boldsymbol{b})$ we have the following constraint that needs to be fulfilled:

$$\sum_A \frac{e^{\beta W(A,b)}}{Z(\boldsymbol{A})} P(\boldsymbol{A}) = P(\boldsymbol{b}). \tag{11}$$

Therefore, not all choices of $P(\boldsymbol{A})$ and $P(\boldsymbol{b})$ are compatible with the posterior distribution and the exact possibilities will depend on the actual shape of

$W(A, b)$. However, one choice that is always possible is a maximum-entropy one,

$$P(A) = \frac{Z(A)}{\Xi}, \qquad P(b) = \frac{\Omega(b)}{\Xi}, \tag{12}$$

with $\Omega(b) = \sum_A e^{\beta W(A,b)}$ and $\Xi = \sum_{A,b} e^{\beta W(A,b)}$. Taking this choice leads to the effective generative model

$$P(A|b) = \frac{e^{\beta W(A,b)}}{\Omega(b)}. \tag{13}$$

Therefore, inferentially interpreting a community detection algorithm with a quality function $W(A, b)$ is equivalent to assuming the generative model $P(A|b)$ and prior $P(b)$ of Eqs. 13 and 12 above. Furthermore, this also means that any arbitrary community detection algorithm implies a description length given (in nats) by[15]

$$\Sigma(A, b) = -\beta W(A, b) + \ln \sum_{A',b'} e^{\beta W(A',b')}. \tag{14}$$

What the preceding results show is that there is no such thing as a "model-free" community detection method, since they are all equivalent to the inference of *some* generative model. The only difference to a direct inferential method is that in that case the modelling assumptions are made explicitly, inviting rather than preventing scrutiny. Most often, the effective model and prior that are equivalent to an *ad hoc* community detection method will be difficult to interpret, justify, or even compute (in general, Eq. 14 cannot be written in closed form).

Furthermore there is no guarantee that the obtained description length of Eq. 14 will yield a competitive or even meaningful compression. In particular, there is no guarantee that this effective inference will not overfit the data. Although we mentioned in Section 2.3 that inference and compression are equivalent, the compression achieved when considering a particular generative model is constrained by the assumptions encoded in its likelihood and prior. If these are poorly chosen, no actual compression might be achieved, for example when comparing to the one obtained with a maximally random model. This is precisely what happens with descriptive community detection methods: they overfit because their implicit modelling assumptions do not accommodate the possibility that a network may be maximally random, or contain a balanced mixture of structure and randomness.

Since we can always interpret any community detection method as inferential, is it still meaningful to categorize some methods as descriptive? Arguably

---

[15] The description length of Eq. 14 is only valid if there are no further parameters in the quality function $W(A, b)$ other than b that are being optimized.

yes, because directly inferential approaches make their generative models and priors explicit, while for a descriptive method we need to extract them from reverse engineering. Explicit modelling allows us to make judicious choices about the model and prior that reflect the kinds of structures we want to detect, relevant scales or lack thereof, and many other aspects that improve their performance in practice, and our understanding of the results. With implicit assumptions we are "flying blind," relying substantially on serendipity and trial-and-error — not always with great success.

It is not uncommon to find criticisms of inferential methods due to a perceived implausibility of the generative models used — such as the conditional independence of the placement of the edges present in the SBM [8] — although these assumptions are also present, but only *implicitly*, in other methods, like modularity maximization (see Sec. 4.1). We discuss this issue further in Sec. 4.8.

The above inferential interpretation is not specific to community detection, but is in fact valid for any learning problem. The set of explicit or implicit assumptions that must come with any learning algorithm is called an "inductive bias." An algorithm is expected to function optimally only if its inductive bias agrees with the actual instances of the problems encountered. It is important to emphasize that no algorithm can be free of an inductive bias, we can only chose *which* intrinsic assumptions we make about how likely we are to encounter a particular kind of data, not *whether* we are making an assumption. Therefore, it is particularly problematic when a method does not articulate explicitly what these assumptions are, since even if they are hidden from view, they exist nonetheless, and still need to be scrutinized and justified. This means we should be particularly skeptical of the impossible claim that a learning method is "model-free," since this denomination is more likely to signal an inability or unwillingness to expose the underlying modelling assumptions, which could potentially be revealed as unappealing and fragile when eventually forced to come under scrutiny.

2.6 Caveats and challenges with inferential methods

Inferential community detection is a challenging task, and is not without its caveats. One aspect they share with descriptive approaches is algorithmic complexity (see Sec. 4.9), and the fact that they in general try to solve NP-hard problems. This means that there is no known algorithm that is guaranteed to produce exact results in a reasonable amount of time, except for very small networks. That does not mean that every instance of the problem is hard to answer, in fact it can be shown that in key cases robust answers can be obtained [45], but in general all existing methods are approximative, with the usual trade-off between accuracy and speed. The quest for general approaches that behave well while being efficient is still ongoing and is unlikely to exhausted soon.

Furthermore, employing statistical inference is not a "silver bullet" that automatically solves every problem. If our models are "misspecified," i.e. represent very poorly the structure present in the data, then our inferences using them will be very limited and potentially misleading (see Sec. 4.8) — the most we can expect from our methodology in this case is to obtain good diagnostics of when this is happening [26]. There is also a typical trade-off between realism and simplicity, such that models that more closely match reality are more difficult to express in simple terms with tractable models. Usually, the more complex a model is, the more difficult becomes its inference. The technical task of using algorithms such as Markov chain Monte Carlo (MCMC) to produce reliable inferences for a complex model is nontrivial and requires substantial expertise, and is likely to be a long-living field of research.

In general it can be said that, although statistical inference does not provide automatic answers, it gives us an invaluable platform where the questions can be formulated more clearly, and allows us to navigate the space of answers using more robust methods and theory.

3 Modularity maximization considered harmful

The most widespread method for community detection is modularity maximization [46], which happens also to be one the most problematic. This method is based on the modularity function,

$$Q(\boldsymbol{A}, \boldsymbol{b}) = \frac{1}{2E} \sum_{ij} \left(A_{ij} - \frac{k_i k_j}{2E} \right) \delta_{b_i, b_j}, \tag{15}$$

where $A_{ij} \in \{0, 1\}$ is an entry of the adjacency matrix, $k_i = \sum_j A_{ij}$ is the degree of node i, b_i is the group membership of node i, and E is the total number of edges. The method consists in finding the partition $\hat{\boldsymbol{b}}$ that maximizes $Q(\boldsymbol{A}, \boldsymbol{b})$,

$$\hat{\boldsymbol{b}} = \underset{\boldsymbol{b}}{\operatorname{argmax}} \, Q(\boldsymbol{A}, \boldsymbol{b}). \tag{16}$$

The motivation behind the modularity function is that it compares the existence of an edge (i, j) to the probability of it existing according to a null model, $P_{ij} = k_i k_j / 2E$, namely that of the configuration model [57] (or more precisely, the Chung-Lu model [58]). The motivation for this method is that we should consider a partition of the network meaningful if the occurrence of edges between nodes of the same group exceeds what we would expect with a random null model without communities.

Despite its widespread adoption, this approach suffers from a variety of serious conceptual and practical flaws, which have been documented extensively [1, 2, 59–61]. The most problematic one is that it *purports* to use an inferential criterion — a deviation from a null generative model — but is in fact merely descriptive. As has been recognized very early, this method categorically fails in its own stated goal, since it always finds high-scoring partitions in networks sampled from its own null model [59]. Indeed, the generative model we used in Fig. 2(a) is exactly the null model considered in the modularity function, which if maximized yields the partition seen in Fig. 2(a). As we already discussed, this result bears no relevance to the underlying generative process, and overfits the data.

The reason for this failure is that the method does not take into account the deviation from the null model in a statistically consistent manner. The modularity function is just a re-scaled version of the assortativity coefficient [62], a correlation measure of the community assignments seen at the endpoints of edges in the network. We should expect such a correlation value to be close to zero for a partition that is determined *before* the edges of the network are placed according to the null model, or equivalently, for a partition chosen at random. However, it is quite a different matter to find a partition that *optimizes* the value of $Q(\boldsymbol{A}, \boldsymbol{b})$, after the network is observed. The deviation from a null model computed in Eq. 15 completely ignores the optimization step of Eq. 16, although it is a crucial part of the algorithm. As a result, the method of modularity maximization tends to massively overfit, and find spurious communities even in networks sampled from its null model. If we search for patterns of correlations in a random graph, most of the time we will find them. This is a pitfall known as "data dredging" or "*p*-hacking," where one searches exhaustively for different patterns in the same data and reports only those that are deemed significant, according to a criterion that does not take into account the fact that we are doing this search in the first place.

We demonstrate this problem in Fig. 6, where we show the distribution of modularity values obtained with a uniform configuration model with $k_i = 5$ for every node i, considering both a random partition and the one that maximizes $Q(\boldsymbol{A}, \boldsymbol{b})$. While for a random partition we find what we would expect, i.e. a value of $Q(\boldsymbol{A}, \boldsymbol{b})$ close to zero, for the optimized partition the value is substantially larger. Inspecting the optimized partition in Fig. 6(c), we see that it corresponds indeed to 15 seemingly clear assortative communities — which by construction bear no relevance to how the network was generated. They have been dredged out of randomness by the optimization procedure.

Somewhat paradoxically, another problem with modularity maximization is that in addition to systematically overfitting, it also systematically *underfits*.

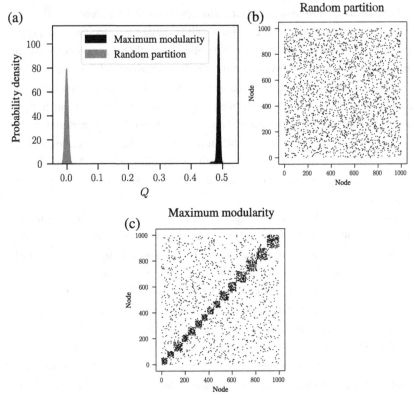

Figure 6 Modularity maximization systematically overfits, and finds spurious structures even its own null model. In this example we consider a random network model with $N = 10^3$ nodes, with every node having degree 5. (a) Distribution of modularity values for a partition into 15 groups chosen at random, and for the optimized value of modularity, for 5000 networks sampled from the same model. (b) Adjacency matrix of a sample from the model, with the nodes ordered according to a random partition. (c) Same as (b), but with the nodes ordered according to the partition that maximizes modularity.

This occurs via the so-called *resolution limit*: in a connected network[16] the method cannot find more than $\sqrt{2E}$ communities [60], even if they seem intuitive or can be found by other methods. An example of this is shown in Fig. 7, where for a network generated with the SBM containing 30 communities,

---

16 Modularity maximization, like many descriptive community detection methods, will always place connected components in different communities. This is another clear distinction with inferential approaches, since maximally random models — without latent community structure — can generate disconnected networks if they are sufficiently sparse. From an inferential point of view, it is therefore incorrect to assume that every connected component must belong to a different community.

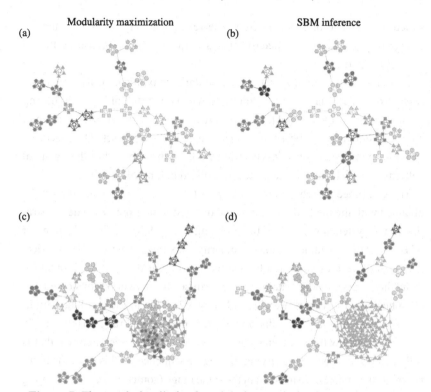

Figure 7 The resolution limit of modularity maximization prevents small communities from being identified, even if there is sufficient statistical evidence to support them. Panel (a) shows a network with $B = 30$ communities sampled from an assortative SBM parametrization. The colors indicate the 18 communities found with modularity maximization, where several pairs of true communities are merged together. Panel (b) shows the inference result of an assortative SBM [24], recovering the true communities with perfect accuracy. Panels (c) and (d) show the results for a similar model where a larger community has been introduced. In (c) we see the results of modularity maximization, which not only merges the smaller communities together, but also splits the larger community into several spurious ones — thus both underfitting and overfitting different parts of the network at the same time. In (d) we see the result obtained by inferring the SBM, which once again finds the correct answer.

modularity maximization finds only 18, while an inferential approach has no problems finding the true structure. There are attempts to counteract the resolution limit by introducing a "resolution parameter" to the modularity function, but as we discuss in Sec. 4.4 they are in general ineffective.

These two problems — overfitting and underfitting — can occur in tandem, such that portions of the network dominated by randomness are spuriously revealed to contain communities, whereas other portions with clear modular structure can have those obstructed. The result is a very unreliable method to capture the structure of heterogeneous networks. We demonstrate this in Fig. 7(c) and (d).

In addition to these major problems, modularity maximization also often possesses a degenerate landscape of solutions, with very different partitions having similar values of $Q(A, b)$ [61]. In these situations the partition with maximum value of modularity can be a poor representative of the entire set of high-scoring solutions and depend on idiosyncratic details of the data rather than general patterns — which can be interpreted as a different kind of overfitting.[17]

The combined effects of underfitting and overfitting can make the results obtained with the method unreliable and difficult to interpret. As a demonstration of the systematic nature of the problem, in Fig. 8(a) we show the number of communities obtained using modularity maximization for 263 empirical networks of various sizes and belonging to different domains [64], obtained from the Netzschleuder catalogue [65]. Since the networks considered are all connected, the values are always below $\sqrt{2E}$, due to the resolution limit; but otherwise they are well distributed over the allowed range. However, in Fig. 8(b) we show the same analysis, but for a version of each network that is fully randomized, while preserving the degree sequence. In this case, the number of groups remains distributed in the same range (sometimes even exceeding the resolution limit, because the randomized versions can end up disconnected). As Fig. 8(c) shows, the number of groups found for the randomized networks is strongly correlated with the original ones, despite the fact that the former have no latent community structure. This is a strong indication of the substantial amount of noise that is incorporated into the partitions found with the method.

The systematic overfitting of modularity maximization — as well as other descriptive methods such as Infomap — has been also demonstrated recently in Ref. [66], from the point of view of edge prediction, on a separate empirical dataset of 572 networks from various domains.

Although many of the problems with modularity maximization were long known, for some time there were no principled solutions to them, but this is no longer the case. In the table below we summarize some of the main problems with modularity and how they are solved with inferential approaches.

---

[17] This kind of degeneracy in the solution landscape can also occur in an inferential setting [26, 63]. However, there it can be interpreted as the existence of competing hypotheses for the same data, whose relative plausibility can be quantitatively assessed via their posterior probability. In case the multiplicity of alternative hypotheses is too large, this would be indicative of poor fit, or a misspecification of the model, i.e. a general inadequacy of the model structure to capture the structure in the data for any possible choice of parameters.

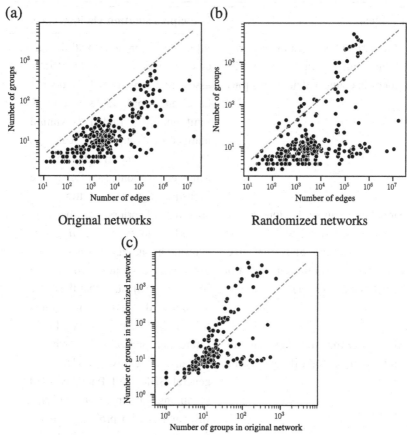

Figure 8 Modularity maximization incorporates a substantial amount of noise into its results. (a) Number of groups found using modularity maximization for 263 empirical networks as a function of the number of edges. The dashed line corresponds to the $\sqrt{2E}$ upper bound due to the resolution limit. (b) The same as in (a) but with randomized versions of each network. (c) Correspondence between the number of groups of the original and randomized network. The dashed line shows the diagonal.

Problem	Principled solution via inference
Modularity maximization overfits, and finds modules in maximally random networks. [59]	Bayesian inference of the SBM is designed from the ground up to avoid this problem in a principled way and systematically succeeds [5].

Problem	Principled solution via inference
Modularity maximization has a resolution limit, and finds at most $\sqrt{2E}$ groups in connected networks [60].	Inferential approaches with hierarchical priors [16, 67] or strictly assortative structures [24] do not have any appreciable resolution limit, and can find a maximum number of groups that scales as $O(N/\log N)$. Importantly, this is achieved without sacrificing the robustness against overfitting.
Modularity maximization has a characteristic scale, and tends to find communities of similar size; in particular with the same sum of degrees (see Sec. 4.4).	Hierarchical priors can be specifically chosen to be *a priori* agnostic about characteristic sizes, densities of groups and degree sequences [16], such that these are not imposed, but instead obtained from inference, in an unbiased way.
Modularity maximization can only find strictly assortative communities.	Inferential approaches can be based on any generative model. The general SBM will find any kind of mixing pattern in an unbiased way, and has no problems identifying modular structure in bipartite networks, core-periphery networks, and any mixture of these or other patterns. There are also specialized versions for bipartite [68], core-periphery [69], and assortative patterns [24], if these are being searched exclusively.
The solution landscape of modularity maximization is often degenerate, with many different solutions with close to the same modularity value [61], and with no clear way of how to select between them.	Inferential methods are characterized by a posterior distribution of partitions. The consensus or dissensus between the different solutions [26] can be used to determine how many cohesive hypotheses can be extracted from inference, and to what extent is the model being used a poor or a good fit for the network.

Because of the above problems, the use of modularity maximization should be discouraged, since it is demonstrably not fit for purpose as an inferential method. As a consequence, the use of modularity maximization in any recent network analysis that relies on inferential conclusions can be arguably considered a "red flag" that strongly indicates methodological inappropriateness. In the absence of secondary evidence supporting the alleged community structures found, or extreme care to counteract the several limitations of the method (see Secs. 4.2, 4.3 and 4.4 for how typical attempts usually fail), the safest assumption is that the results obtained with that method tend to contain a substantial amount of noise, rendering any inferential conclusion derived from them highly suspicious.

As a final note, we focus on modularity here not only for its widespread adoption but also because of its exemplary character. At a fundamental level, all of its shortcoming are shared with any descriptive method in the literature — to varied but always non-negligible degrees.

4 Myths, pitfalls, and half-truths

In this section we focus on assumed or asserted statements about how to circumvent pitfalls in community detection, which are in fact better characterized as myths or half-truths, since they are either misleading, or obstruct a more careful assessment of the true underlying nature of the problem. The following subsections each deal with one of these pitfalls.

4.1 "Modularity maximization and SBM inference are equivalent methods."

As we have discussed in Sec. 2.5, it is possible to interpret *any* community detection algorithm as the inference of *some* generative model. Because of this, the mere fact that an equivalence with an inferential approach exists cannot be used to justify the inferential use of a descriptive method, or to use it as a criterion to distinguish between approaches that are statistically principled or not. To this aim, we need to ask instead whether the modelling assumptions that are *implicit* in the descriptive approach can be meaningfully justified, and whether they can be used to consistently infer structures from networks.

Some recent works have detailed some specific equivalences of modularity maximization with statistical inference [70, 71]. As we will discuss in this section, these equivalences are far more limited than commonly interpreted. They serve mostly to understand in more detail the reasons why modularity maximization fails as a reliable method, but do not prevent it from failing — they expose more clearly its sins, but offer no redemption.

We start with a very interesting connection revealed by Zhang and Moore [70] between the effective posterior distribution we obtain when using the modularity function as a Hamiltonian,

$$P(b|A) = \frac{e^{\beta E Q(A,b)}}{Z(A)}, \tag{17}$$

and the posterior distribution of the strictly assortative DC-SBM, which we refer here as the degree-corrected planted partition model (DC-PP),

$$P(b|A, \omega_{in}, \omega_{out}, \theta) = \frac{P(A|\omega_{in}, \omega_{out}, \theta, b)P(b)}{P(A|\omega_{in}, \omega_{out}, \theta)}, \tag{18}$$

which has a likelihood given by

$$P(A|\omega_{in}, \omega_{out}, \theta, b) = \prod_{i<j} \frac{e^{-\omega_{b_i,b_j}\theta_i\theta_j}\left(\omega_{b_i,b_j}\theta_i\theta_j\right)^{A_{ij}}}{A_{ij}!}, \tag{19}$$

where

$$\omega_{rs} = \omega_{in}\delta_{rs} + \omega_{out}(1-\delta_{rs}). \tag{20}$$

This model assumes that there are constant rates ω_{in} and ω_{out} controlling the number of edges that connect to nodes of the same and different communities, respectively. In addition, each node has its own propensity θ_i, which determines the relative probability it has of receiving an edge, such that nodes inside the same community are allowed to have very different degrees. This is a far more restrictive version of the full DC-SBM we considered before, since it not only assumes assortativity as the only mixing pattern, but also that all communities share the same rate ω_{in}, which imposes a rather unrealistic similarity between the different groups.

Before continuing, it is important to emphasize that the posterior of Eq. 18 corresponds to the situation where the number of communities and all parameters of the model, except the partition itself, are known *a priori*. This does not correspond to any typical empirical setting where community detection is employed, since we do not often have such detailed information about the community structure, and in fact no good reason to even use this particular parametrization to begin with. The equivalences that we are about to consider apply only in very idealized scenarios, and are not expected to hold in practice.

Taking the logarithm of both sides of Eq. 19, and ignoring constant terms with respect to the model parameters we have

$$\ln P(A|\omega_{\text{in}}, \omega_{\text{out}}, \boldsymbol{\theta}, \boldsymbol{b}) = \left(\ln \frac{\omega_{\text{in}}}{\omega_{\text{out}}}\right) \left[\sum_{i<j} \left(A_{ij} - \frac{\omega_{\text{in}} - \omega_{\text{out}}}{\ln(\omega_{\text{in}}/\omega_{\text{out}})} \theta_i \theta_j\right) \delta_{b_i, b_j}\right] +$$
$$\sum_{i<j} \left[A_{ij} \ln(\theta_i \theta_j \omega_{\text{out}}) - \theta_i \theta_j \omega_{\text{out}}\right]. \quad (21)$$

Therefore, ignoring additive terms that do not depend on $\boldsymbol{b}$ (since they become irrelevant after normalization in Eq. 17) and making the arbitrary choices (we will inspect these in detail soon),

$$\beta = \ln \frac{\omega_{\text{in}}}{\omega_{\text{out}}}, \qquad \frac{\ln(\omega_{\text{in}}/\omega_{\text{out}})}{\omega_{\text{in}} - \omega_{\text{out}}} = 2E, \qquad \theta_i = k_i, \qquad (22)$$

we obtain the equivalence,

$$\ln P(A|\omega_{\text{in}}, \omega_{\text{out}}, \boldsymbol{\theta}, \boldsymbol{b}) = \beta E Q(A, \boldsymbol{b}), \qquad (23)$$

allowing us to equate Eqs. 17 and 18 (there is a methodological problem with the choice $\theta_i = k_i$, as we will see later, but we will ignore this for the time being). Therefore, for particular choices of the model parameters, one recovers modularity optimization from the maximum likelihood estimation of the DC-PP model with respect to $\boldsymbol{b}$. Indeed, this allows us to understand more clearly what *implicit* assumptions go behind using modularity for inferential aims. For example, besides making very specific prior assumptions about the model parameters ω_{in}, ω_{out} and $\boldsymbol{\theta}$, this posterior also assumes that all partitions are equally likely *a priori*,

$$P(\boldsymbol{b}) \propto 1. \qquad (24)$$

We can in fact write this uniform prior more precisely as

$$P(\boldsymbol{b}) = \left[\sum_{B=1}^{N} \left\{{N \atop B}\right\} B!\right]^{-1}, \qquad (25)$$

where $\left\{{N \atop B}\right\} B!$ is the number of labelled partitions of a set N into B groups. This number reaches a maximum at $B \approx 0.72 \times N$, and decays fast from there, meaning that such a uniform prior is in fact very concentrated on a very large number of groups — partially explaining the tendency of the modularity posterior to overfit. Let us examine now the prior assumption

$$\frac{\ln(\omega_{\text{in}}/\omega_{\text{out}})}{\omega_{\text{in}} - \omega_{\text{out}}} = 2E. \qquad (26)$$

For any value of E, Eq. (26) admits many solutions. However, not all of them are consistent with the *expected* number of edges in the network according to the DC-PP model. Assuming, for simplicity, that all B groups have the same

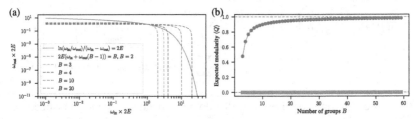

Figure 9 Using modularity maximization is equivalent to performing a maximum likelihood estimate of the DC-PP model with very specific parameter choices, that depend on the number of edges E in the network and the number of communities B. In (a) we show the valid choices of ω_{in} and ω_{out} obtained when the solid and dashed lines cross, corresponding respectively to Eqs. 26 and 28, where we can see that for $B = 2$ no solution is possible where the expected modularity is positive. In (b) we show the two possible values for the expected modularity that are consistent with the implicit model assumptions, as a function of the number of groups.

size N/B, and that all nodes have the same degree $2E/N$, then the expected number of edges according to the assumed DC-PP model is given by

$$2\langle E \rangle = (2E)^2 \left(\frac{\omega_{in}}{B} + \frac{\omega_{out}(B-1)}{B} \right). \tag{27}$$

Equating the expected with the observed value, $\langle E \rangle = E$, leads to

$$\omega_{in} + \omega_{out}(B-1) = \frac{B}{2E}. \tag{28}$$

Combining Eqs. 26 and 28 gives us at most only two values of ω_{in} and ω_{out} that are compatible with the expected density of the network and the modularity interpretation of the likelihood, as seen in Fig. 9(a), and therefore only two possible values for the expected modularity, computed as

$$\langle Q \rangle = \frac{1}{B} \left(2E\omega_{in} - 1 \right). \tag{29}$$

One possible solution is always $\omega_{in} = \omega_{out} = 1/2E$, which leads to $\langle Q \rangle = 0$. The other solution is only possible for $Q > 2$, and yields a specific expected value of modularity which approaches $\langle Q \rangle \to 1$ as B increases (see Fig. 9(b)). This yields an implausibly narrow range for the consistency of modularity maximization with the inference of the DC-PP model. The bias towards larger values of $Q(\mathbf{A}, \mathbf{b})$ as the number of groups increases is not an inherent property of the DC-PP model, as it accommodates any expected value of modularity by properly choosing its parameters. Instead, this is an arbitrary implicit assumption baked in $Q(\mathbf{A}, \mathbf{b})$, which further explains why maximizing it will tend to find many groups even on random networks.

In a later work [71], Newman relaxed the above connection with modularity by using instead its generalized version [72, 73],

$$Q(A, b, \gamma) = \frac{1}{2E} \sum_{ij} \left(A_{ij} - \gamma \frac{k_i k_j}{2E} \right) \delta_{b_i, b_j}, \tag{30}$$

where γ is the so-called "resolution" parameter. With this additional parameter, we have more freedom about the implicit assumptions of the DC-PP model. Newman in fact showed that if we make the choices,

$$\beta = \ln \frac{\omega_{\text{in}}}{\omega_{\text{out}}}, \qquad \gamma = \frac{\omega_{\text{in}} - \omega_{\text{out}}}{\ln(\omega_{\text{in}}/\omega_{\text{out}})}, \qquad \theta_i = \frac{k_i}{\sqrt{2E}}, \tag{31}$$

then we recover the Gibbs distribution with the generalized modularity from the DC-PP likelihood of Eq. 19. Due to the independent parameter γ, now the assumed values of ω_{in} and ω_{out} are no longer constrained by E alone, and can take any value. Therefore, if we knew the correct value of the model parameters, we could use them to choose the appropriate value of γ and hence maximize $Q(A, b, \gamma)$, yielding the same answer as maximizing $\ln P(A|\omega_{\text{in}}, \omega_{\text{out}}, \theta, b)$ with the same parameters.

There are, however, serious problems remaining that prevent this equivalence from being true in general, or in fact even typically. For the equivalence to hold, we need the number of groups B and all parameters to be known a priori and to be equal to Eq. 31. However, the choice $\theta_i = k_i/\sqrt{2E}$ involves information about the observed network, namely the actual degrees seen — and therefore is not just a prior assumption, but one done *a posteriori*, and hence needs to be justified via a consistent estimation that respects the likelihood principle. When we perform a maximum likelihood estimate of the parameters ω_{in}, ω_{out}, and θ, we obtain the following system of nonlinear equations [24],

$$\omega_{\text{in}}^* = \frac{2e_{\text{in}}}{\sum_r \hat{\theta}_r^2} \tag{32}$$

$$\omega_{\text{out}}^* = \frac{e_{\text{out}}}{\sum_{r<s} \hat{\theta}_r \hat{\theta}_s} \tag{33}$$

$$\theta_i^* = k_i \left[\frac{2e_{\text{in}} \hat{\theta}_{b_i}}{\sum_r \hat{\theta}_r^2} + \frac{e_{\text{out}} \sum_{r \neq b_i} \hat{\theta}_r}{\sum_{r<s} \hat{\theta}_r \hat{\theta}_s} \right]^{-1}, \tag{34}$$

where $e_{\text{in}} = \sum_{i<j} A_{ij} \delta_{b_i, b_j}$, $e_{\text{out}} = \sum_{i<j} A_{ij}(1 - \delta_{b_i, b_j})$, and $\hat{\theta}_r = \sum_i \theta_i^* \delta_{b_i, r}$. This system (Eqs. (32), (33), and (34)) admits $\theta_i^* = k_i/\sqrt{2E}$ as a solution only if the following condition is met for every group r:

$$\sum_i k_i \delta_{b_i, r} = \frac{2E}{B}. \tag{35}$$

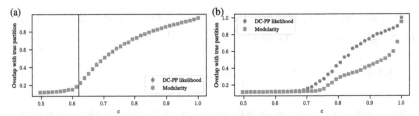

Figure 10 Generalized modularity and the DC-PP model are only equivalent if the symmetry of Eq. 35 is preserved. Here we consider an instance of the DC-PP model with $\omega_{\text{in}} = 2Ec/N$, $\omega_{\text{out}} = 2E(1 - c)/\sum_{r \neq s} \sqrt{n_r n_s}$, and $\theta_i = 1/\sqrt{n_{b_i}}$, where n_r is the number of nodes in group r. The parameter $c \in [0, 1]$ controls the degree of assortativity. For non-uniform group sizes, the symmetry of Eq. 35 is not preserved with this choice of parameters. We use the parametrization $n_r = N\alpha^{r-1}(1 - \alpha)/(1 - \alpha^B)$, where $\alpha > 0$ controls the group size heterogeneity. When employing generalized modularity, we choose the closest possible parameter choice with $\omega_{\text{in}} = 2Ec/(\sum_r e_r^2/2E)$ and $\omega_{\text{out}} = 2E(1 - c)/(2E - \sum_r e_r^2/2E)$, where $e_r = \sum_i k_i \delta_{b_i,r}$. In (a) we show the inference results for the uniform case with $\alpha \to 1$, where both approaches are identical, performing equally well all the way down to the detectability threshold [6] (vertical line). In (b) we show the result with $\alpha = 2$, which leads to unequal group sizes, causing the behavior between both approaches to diverge. In all cases we consider averages over 5 networks with $N = 10^4$ nodes, average degree $2E/N = 3$, and $B = 10$ groups.

In other words, the sum of degrees inside each group must be the same for every group. Note also that the expected degrees according to the DC-PP model will be inconsistent with Eq. 31 if the condition in Eq. (35) is not met, i.e.

$$\langle k_i \rangle = \theta_i \left[\omega_{\text{in}} \sum_j \theta_j \delta_{b_i,b_j} + \omega_{\text{out}} \sum_j \theta_j (1 - \delta_{b_i,b_j}) \right]. \tag{36}$$

Substituting $\theta_i = k_i/\sqrt{2E}$ in the above equation will yield in general $\langle k_i \rangle \neq k_i$, as long as Eq. 35 is not fulfilled, *regardless of how we choose ω_{in} and ω_{out}*.

Framing it differently, for any choice of ω_{in}, ω_{out} and $\boldsymbol{\theta}$ such that the sums $\sum_i \theta_i \delta_{b_i,r}$ are not identical for every group r, the DC-SBM likelihood $\ln P(\boldsymbol{A}|\omega_{\text{in}}, \omega_{\text{out}}, \boldsymbol{\theta}, \boldsymbol{b})$ is not captured by $Q(\boldsymbol{A}, \boldsymbol{b}, \gamma)$ for any value of γ, and therefore maximizing both functions will not yield the same results. That is, the equivalence is only valid for special cases of the model *and* data. We show in Fig. 10 an example of an instance of the DC-PP model where the generalized modularity yields results which are inconsistent with using likelihood of the DC-PP model directly.

Because of the above caveats, we have to treat the claimed equivalence with a grain of salt. In general there are only three scenarios we may consider when analysing a network:

1. We know that the network has been sampled from the DC-PP model, as well as the correct number of groups B and the values of the parameters ω_{in}, ω_{out}, and θ, and the following symmetry exists:

$$\sum_i \theta_i \delta_{b_i,r} = C, \tag{37}$$

where C is a constant.
2. Like the first case, but where the symmetry of Eq. 37 does not exist.
3. Every other situation.

Cases 1 and 2 are highly idealized and are not expected to be encountered in practice, which almost always falls in case 3. Nevertheless, the equivalence between the DC-PP model and generalized modularity is only valid in case 1. In case 2, as we already discussed, the use of generalized modularity will be equivalent to *some* generative model — as all methods are — but which cannot be expressed within the DC-PP parametrization.

Because of the above problems, the relevance of this partial equivalence between these approaches in practical scenarios is arguably dubious. It serves only to demonstrate how the implicit assumptions behind modularity maximization are hard to justify.

We emphasize also the obvious fact that even if the equivalence with the DC-PP model were to hold more broadly, this would not make the pathological behavior of modularity described in Sec. 3 disappear. Instead, it would only show that this particular inferential method would *also* be pathological. In fact, it is well understood that maximum likelihood is not in general an appropriate inferential approach for models with an arbitrarily large number of degrees of freedom, since it lacks the regularization properties of Bayesian methods [5], such as the one we described in Sec. 2.2, where instead of considering point estimates of the parameters, we integrate over all possibilities, weighted according to their prior probability. In this way, it is possible to *infer* the number of communities, instead of assuming it *a priori*, together with all other model parameters. In fact, when such a Bayesian approach is employed for the DC-PP model, one obtains the following marginal likelihood [24],

$$P(A|b) = \int P(A|\omega_{\text{in}}, \omega_{\text{out}}, \theta, b) P(\omega_{\text{in}}) P(\omega_{\text{out}}) P(\theta|b) \, d\omega_{\text{in}} d\omega_{\text{out}} d\theta \tag{38}$$

$$= \frac{e_{\text{in}}! e_{\text{out}}!}{\left(\frac{B}{2}\right)^{e_{\text{in}}} \binom{B}{2}^{e_{\text{out}}} (E+1)^{1-\delta_{B,1}}} \prod_r \frac{(n_r-1)!}{(e_r+n_r-1)!} \times \frac{\prod_i k_i!}{\prod_{i<j} A_{ij}! \prod_i A_{ii}!!},$$

$$\tag{39}$$

where $e_{\text{in}} = \sum_{i<j} A_{ij}\delta_{b_i,b_j}$ and $e_{\text{out}} = E - e_{\text{in}}$. As demonstrated in Ref. [24], this approach allows us to detect purely assortative community structures in a nonparametric way, in a manner that prevents both overfitting and underfitting — i.e. the resolution limit *vanishes* since we inherently consider every possible value of the parameters ω_{in} and ω_{out} — thus lifting two major limitations of modularity. Note also that Eq. 39 (or its logarithm) does not bear any direct resemblance to the modularity function, and therefore it does not seem possible to reproduce its behavior via a simple modification of the latter.[18]

We also mention briefly a result obtained by Bickel and Chen [74], which states that modularity maximization can consistently identify the community assignments of networks generated by the SBM in the dense limit. This limit corresponds to networks where the average number of neighbors is comparable to the total number of nodes. In this situation, the community detection problem becomes substantially easier, and many algorithms, including e.g. unregularized spectral clustering, can do just as well as modularity maximization. This result tells us more about how easy it is to find communities in dense networks than about the quality of the algorithms compared. The dense scenario does not represent well the difficulty of finding communities in real networks, which are overwhelmingly sparse, with an average degree much smaller than the total number of nodes. In the sparse case, likelihood-based inferential approaches are optimal and outperform modularity [6, 74]. Comparable equivalences have also been encountered with spectral methods [75], but they also rely on particular realizations of the community detection problem, and do not hold in general.

In short, if the objective is to infer the DC-PP model, there is no reason to do it via the maximization of $Q(A, b, \gamma)$, nor is it in general equivalent to any consistent inference approach such as maximum likelihood or Bayesian posterior inference. Even in the unlikely case where the true number of communities is known, the implicit assumptions of modularity correspond to the DC-PP model not only with uniform probabilities between communities but also uniform sums of degrees for every community. If these properties are not present in the network, the method offers no inherent diagnostic, and will find spurious structures that tend to match it, regardless of their statistical significance. Combined with the overall lack of regularization, these features render the method substantially prone to distortion and overfitting. Ultimately, the use of any form of modularity maximization fails the litmus test we considered earlier, and should be considered a purely descriptive community detection

---

[18] There is also no need to "fix" modularity. We can simply use Eq. 39 in its place for most algorithms, which incurs almost no additional computational overhead.

method. Whenever the objective is to understand network structure, it needs to be replaced with a flexible and robust inferential procedure.

4.2 "Consensus clustering can eliminate overfitting."

As mentioned in Sec. 3, methods like modularity maximization tend to have a degenerate solution landscape. One strategy proposed to tackle this problem is to obtain a *consensus clustering*, i.e. leverage the entire landscape of solutions to produce a single partition that points in a cohesive direction, representative of the whole ensemble [63, 76, 77]. If no cohesive direction exists, one could then conclude that no actual community structure exists, and therefore solve the overfitting problem of finding communities in maximally random networks. In reality, however, a descriptive community detection method can in fact display a cohesive set of solutions on a maximally random network. We demonstrate this in Fig. 11 which shows the consensus between 10^5 different maximum modularity solutions for a small random network, using the method of Ref. [26] to obtain the consensus. Although we can notice a significant variability between the different partitions, there is also substantial agreement. In particular, there is no clear indication from the consensus that the underlying network is maximally random. The reason for this that the randomness of the network is *quenched*, and does indeed point to a specific community structure with the highest modularity. The ideas of solution heterogeneity and overfitting are, in general, orthogonal concepts.

With care, it is possible to probe the solution landscape in a manner that reveals a signal of the randomness of the underlying network. For this purpose, some authors have proposed that instead of finding the maximum modularity partition, one instead samples them from the Gibbs distribution [70, 76, 78, 79],

$$P(b) = \frac{e^{\beta Q(A,b)}}{Z(A)}, \tag{40}$$

with normalization $Z(A) = \sum_b e^{\beta Q(A,b)}$, effectively considering $Q(A,b)$ as the Hamiltonian of a spin system with an inverse temperature parameter β. For a sufficiently large random network, there is a particular value $\beta = \beta^*$, below which samples from the distribution become uncorrelated, forming a lack of consensus [70]. There is a problem, however: there is no guarantee that if a lack of consensus exists for $\beta < \beta^*$, then the network must be random; only the reverse is true. In general, while statements can be made about the behavior of the modularity landscape for maximally random and sufficiently large networks, or even for networks sampled from a SBM, very little can be said about its behavior on real, finite networks. Since real networks are likely to

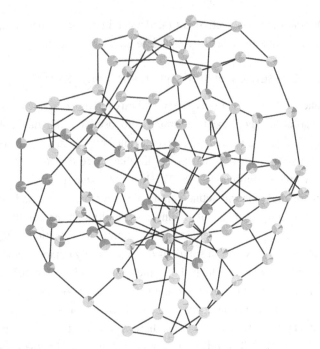

Figure 11 Consensus clustering of a maximally random network, sampled from the Erdő-Rényi model, that combines 10^5 solutions of the maximum modularity method. On each node there is a pie chart describing the frequencies with which it was observed in a given community, obtained using the approach described in Ref. [26]. Despite the lack of latent communities, there is a substantial agreement between the different answers.

contain a heterogeneous mixture of randomness and structure (e.g. as illustrated in Fig. 7(c)) this kind of approach becomes ultimately unreliable. One fundamental problem here is that these approaches attempt to reach an inferential conclusion ("is the network sampled from a random model?") without fully going through Bayes' formula of Eq. 1, and reasoning about model assumptions, prior information and compressibility. We currently lack a principled methodology to reach such a conclusion while avoiding these crucial steps.

Another aspect of the relationship between consensus clustering and overfitting is worth mentioning. In an inferential setting, if we wish to obtain an estimator for the true partition $\hat{b}$, this will in general depend on how we evaluate its accuracy. In other words, we must define an error function $\varepsilon(b', b)$ such that

$$b = \underset{b'}{\arg\min}\ \varepsilon(b', b). \tag{41}$$

Based on this, our best possible estimate is the one which minimizes the average error over the entire posterior distribution,

$$\hat{b} = \underset{b'}{\mathrm{argmin}} \sum_{b} \varepsilon(b', b) P(b|A). \tag{42}$$

Note that in general this estimator will be different from the most likely partition, i.e.

$$\hat{b} \neq \underset{b}{\mathrm{argmax}} \, P(b|A). \tag{43}$$

The optimal estimator $\hat{b}$ will indeed correspond to a consensus over all possible partitions, weighted according to their plausibility. In situations where the posterior distribution is concentrated on a single partition, both estimators will coincide. Otherwise, the most likely partition might in fact be less accurate and incorporate more noise than the consensus estimator, which might be seen as a form of overfitting. This kind of overfitting is of a different nature than the one we have considered so far, since it amounts to a residual loss of accuracy, where an (often small) fraction of the nodes end up incorrectly classified, instead of spurious groups being identified. However, there are many caveats to this kind of analysis. First, it will be sensitive to the error function chosen, which needs to be carefully justified. Second, there might be no cohesive consensus, in situations where the posterior distribution is composed of several distinct "modes," each corresponding to a different hypothesis for the network. In such a situation the consensus between them might be unrepresentative of the ensemble of solutions. There are principled approaches to deal with this problem, as described in Refs. [26, 80].

4.3 "Overfitting can be tackled by doing a statistical significance test of the quality function."

Sometimes practitioners are aware that non-inferential methods like modularity maximization can find communities in random networks. In an attempt to extract an inferential conclusion from their results, they compare the value of the quality function with a randomized version of the network — and if a significant discrepancy is found, they conclude that the community structure is statistically meaningful [78]. Unfortunately, this approach is as fundamentally flawed as it is straightforward to implement.

The reason why the test fails is because in reality it answers a question that is different from the one intended. When we compare the value of the quality function obtained from a network and its randomized counterpart, we can use this information to answer *only* the following question: "Can we reject the hypothesis that the observed network was sampled from a random null model?" No other information can be obtained from this test, including whether the *network partition* we obtained is significant. All we can determine is if the optimized value of the quality function is significant or not. The distinction between the significance of the quality function value and the network partition itself is subtle but crucial.

We illustrate the above difference with an example in Fig. 12(b). This network is created by starting with a maximally random Erdős-Rényi (ER) network, and adding to it a few more edges so that it has an embedded clique of six nodes. The occurrence of such a clique from an ER model is very unlikely, so if we perform a statistical test on this network that is powerful enough, we should be able to rule out that it came from the ER model with good confidence. Indeed, if we use the value of maximum modularity for this test, and compare with the values obtained for the ER model with the name number of nodes and edges (see Fig. 12(a)), we are able to reach the correct conclusion that the null model should be rejected, since the optimized value of modularity is significantly higher for the observed network. Should we conclude therefore that the communities found in the network are significant? If we inspect Fig. 12(b), we see that the maximum value of modularity indeed corresponds to a more-or-less decent detection of the planted clique. However, it also finds another seven completely spurious communities in the random part of the network. What is happening is clear — the planted clique is enough to increase the value of Q such that it becomes a suitable test to reject the null model,[19] but the test is not able to determine that the communities themselves are statistically meaningful. In short, the statement "the value of Q is significant" is not synonymous with "the network partition is significant." Conflating the two will lead to the wrong conclusion about the significance of the communities uncovered.

In Fig. 12(c) we show the result of a more appropriate inferential approach, based on the SBM as described in Sec. 2.3, that attempts to answer a much more relevant question: "which partition of the network into groups is more likely?" The result is able to cleanly separate the planted clique from the rest of the network, which is grouped into a single community.

This example also shows how the task of rejecting a null model is very oblique to Bayesian inference of generative models. The former attempts to determine what the network *is not*, while the latter what *it is*. The first task

---

[19] Note that it is possible to construct alternative examples, where instead of planting a clique, we introduce the placement of triangles, or other features that are known to increase the value of modularity, but that do not correspond to an actual community structure [81].

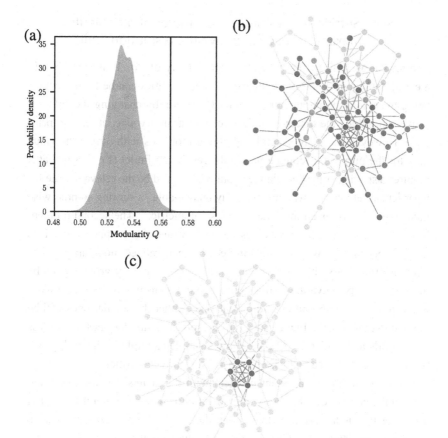

Figure 12 The statistical significance of the maximum modularity value is not informative of the significance of the community structure. In (a) we show the distribution of optimized values of modularity for networks sampled from the Erdős-Rényi (ER) model with the same number of nodes and edges as the network shown in (b) and (c). The vertical line shows the value obtained for the partition shown in (b), indicating that the network is very unlikely to have been sampled from the ER model ($P = 0.002$). However, what sets this network apart from typical samples is the existence of a small clique of six nodes that would not occur in the ER model. The remaining communities found in (b) are entirely meaningless. In (c) we show the result of inferring the SBM on this network, which perfectly identifies the planted clique without overfitting the rest of the network.

tends to be easy — we usually do not need very sophisticated approaches to determine that our data did not come from a null model, specially if our data is complex. On the other hand, the second task is far more revealing, constructive, and arguably more useful in general.

4.4 "Setting the resolution parameter of modularity maximization can remove the resolution limit."

The resolution limit of the generalized modularity of Eq. 30 is such that, in a connected network, no more than $\sqrt{\gamma 2E}$ communities can be found, with γ being the resolution parameter [72, 73]. Therefore, by changing the value of γ, we can induce the discovery of modules of arbitrary size, at least in principle. However, there are several underlying problems with tuning the value of γ for the purpose of counter-acting the resolution limit. The first is that it requires a specific prior knowledge about what would be the relevant scale for a particular network — which is typically unavailable — turning an otherwise nonparametric approach into one which is parametric.[20] The second problem is even more serious: In many cases no single value of γ is appropriate. This happens because, as we have seen in Sec. 4.1, generalized modularity comes with the built-in assumption that the sum of degrees of every group should be the same. The preservation of this homogeneity means that when the network is composed of communities of different sizes, either the smaller ones will be merged together or the bigger ones will be split into smaller ones, regardless of the statistical evidence [82]. We show a simple example of this in Fig. 13, where no value of γ can be used to recover the correct partition.

However, the most important problem with the analysis of the resolution limit in the context of modularity maximization is that it is often discussed in a manner that is largely decoupled from the issue of statistical significance. Since we can interpret a limit on the maximum number of groups as type of systematic underfitting, we can only meaningfully discuss the removal of this limitation if we also do not introduce a tendency to *overfit*, i.e. find more groups than justifiable by statistical evidence. This is precisely the problem with "mutliresolution" approaches [83], or analyses of quality functions other than modularity [84], that claim a reduced or a lack of resolution limit, but without providing a robustness against overfitting. This one-sided evaluation is fundamentally incomplete, as we may end up trading one serious limitation for another.

Methods based on the Bayesian inference of the SBM can tackle the issue of over- and underfitting, as well as preferred sizes of communities at the source. As was shown in Ref. [85], a uninformative assumption about the mixing patterns between groups leads naturally to a resolution limit similar to the one

---

[20] We emphasize that the maximum likelihood approach proposed in Ref. [71] to determine γ, even ignoring the caveats discussed in Sec. 4.1 that render it invalid unless very specific conditions are met, is only applicable for situations when the number of groups is known, directly undermining its use to counteract the resolution limit.

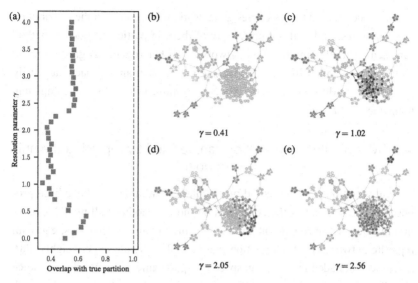

Figure 13 Modularity maximization imposes characteristic community sizes in a manner that hides heterogeneity. Panel (a) shows the overlap between the true and obtained partition for the network described in Fig. 7, as a function of the resolution parameter γ. Panels (b) to (e) show the partitions found for different values of γ, where we see that as smaller groups are uncovered, bigger ones are spuriously split. The result is that no value of γ allows the true communities to be uncovered.

existing for modularity, where no more than $O(\sqrt{N})$ groups can be inferred for sparse networks. However, since in an inferential context our assumptions are made explicitly, we can analyse them more easily and come up with more appropriate choices. In Ref. [67] it was shown how replacing the noninformative assumption by a Bayesian hierarchical model can essentially remove the resolution limit, with a maximum number of groups scaling as $O(N/\log N)$. That model is still unbiased with respect to the expected mixing patterns, and incorporates only the assumption that the patterns themselves are generated by another SBM, with its own patterns generated by yet another SBM, and so on recursively. Another model that has also been shown to be free of the resolution limit is the assortative SBM of Ref. [24]. Importantly, in both these cases the removal of the resolution limit is achieved without sacrificing the capacity of the method to avoid overfitting — e.g. none of these approaches will find spurious groups in random networks.

The issue with preferred group sizes can also be tackled in a principled way in an inferential setting. As demonstrated in Ref. [16], we can also design Bayesian prior hierarchies where the group size distribution is chosen in a noninformative manner, before the partition itself is determined. This results in an inference method that is by design agnostic with respect to the distribution of group sizes, and will not prefer any of them in particular. Such a method can then be safely employed on networks with heterogeneous group sizes in an unbiased manner. In Fig. 7(d) we show how such an approach can easily infer groups of different sizes for the same example of Fig. 13, in a completely nonparametric manner.

4.5 "Modularity maximization can be fixed by replacing the null model."

Several variations of the method of modularity maximization have been proposed, where instead of the configuration model, another null model is used, in a manner that makes the method applicable in various scenarios, e.g. with bipartite networks [86], correlation matrices [87], signed edge weights [88], networks embedded in euclidean spaces [89], to name a few. While the choice of null model has an important effect on what kind of structures are uncovered, its choice does not address any of the statistical shortcomings of modularity that we consider here. In general, just like it happens for the configuration model, the approach will find spurious communities in networks sampled from its null model, regardless of how it is chosen. As as discussed in Sec. 3, this happens because the measured deviation does not account for the optimization procedure employed. Any method based on optimizing the modularity score will amount to a data dredging procedure, independently of the null model chosen, and are thus unsuitable for inferential aims.

4.6 "Descriptive approaches are good enough when the community structure is obvious."

A common argument goes that, sometimes, the community structure of a network is so "obvious" that it will survive whatever abuse we direct at it, and it will be uncovered by a majority of community detection methods that we employ. Therefore, if we are confident that our network contains a clear signal of its community structure, specially if several algorithms substantially agree with each other, or they agree with metadata, then it does not matter very much which algorithm we use.

There are several problems with this argument. First, if an "obvious" structure exists, it does not necessarily mean that it is really meaningful, or statistically significant. If ten algorithms overfit, and one does not, the majority vote is incorrect, and we should prefer the minority opinion. This is precisely the case we considered in Fig. 2, where virtually any descriptive method would uncover the same 13 communities — thus overfitting the network — while an inferential approach would not. And if a method agrees with metadata, while another finds further structure not in agreement, what is to say that this structure is not really there? (Metadata are not "ground truth," they are only more data [90–92], and hence can have its own complex, incomplete, noisy, or even irrelevant relationship with the network.)

Secondly, and even more importantly, how do we even define what is an "obvious" community structure? In general, networks are not low dimensional objects, and we lack methods to inspect their structure directly, a fact which largely motivates community detection in the first place. Positing that we can just immediately determine community structure largely undermines this fact. Often, structure which is deemed "obvious" at first glance, ceases to be so upon closer inspection. For example, one can find claims in the literature that different connected components must "obviously" correspond to different communities. However, maximally random graphs can end up disconnected if they are sufficiently sparse, which means that from an inferential point of view different components can belong to the same community.

Another problem is that analyses of community detection results rely frequently on visual inspections of graphical network layouts, where one tries to evaluate if the community labels agree with the position of the nodes. However, the positioning of the nodes and edges is not inherent to the network itself, and needs to be obtained with some graph drawing algorithm. A typical example are the so-called "spring-block" or "force-directed" layouts, where one considers attractive forces between nodes connected by an edge (like a spring) and an overall repulsive force between all nodes [93]. The final layout is then obtained by minimizing the energy of the system, resulting in edges that have similar length and as few crossings between edges as possible (e.g. in Fig. 1 we used the algorithm of Ref. [93]). This kind of drawing in itself can be seen as a type of indirect descriptive community detection method, since nodes belonging to the same assortative community will tend to be placed close to each other [94]. Based on this observation, when we say that we "see" the communities in a drawing like in Fig. 1, we are in reality only seeing what the layout algorithm is telling us. Therefore, we should always be careful when comparing the results we get with a community detection algorithm to the structure we see in these layouts, because there is no reason to assume that the layout algorithm itself is doing a better job than the clustering algorithm we are evaluating.[21] In fact, this

---

[21] Indeed, if we inspect Fig. 11, which shows the consensus clustering of a maximally random network, we notice that nodes that are classified in the same community end up close together in the drawing, i.e. the layout algorithm also agrees with the modularity consensus. Therefore, it should not be used as a "confirmation" of the structure any more than the result of any other community detection algorithm, since it is also overfitting from an inferential perspective.

is often not the case, since the actual community structures in many networks do not necessarily have a sufficiently low-dimensional representation that is required for this kind of visualization to be effective.

4.7 "The no-free-lunch theorem means that every community detection method is equally good."

For a wide class of optimization and learning problems there exist so-called "no-free-lunch" (NFL) theorems, which broadly state that when averaged over all possible problem instances, all algorithms show equivalent performance [95–97]. Peel *et al* [92] have proved that this is also valid for the problem of community detection, meaning that no single method can perform systematically better than any other, when averaged over all community detection problems. This has been occasionally interpreted as a reason to reject the claim that we should systematically prefer certain classes of algorithms over others. This is, however, a misinterpretation of the theorem, as we will now discuss.

The NFL theorem for community detection is easy to state. Let us consider a generic deterministic community detection algorithm indexed by f, defined by the function $\hat{b}_f(A)$, which ascribes a single partition to a network A. Peel *et al* [92] consider an instance of the community detection problem to be an arbitrary pair (A, b) composed of a network A and the correct partition b that one wants to find from A. We can evaluate the accuracy of the algorithm f via an error (or "loss") function

$$\varepsilon(b, \hat{b}_f(A)), \tag{44}$$

which should take the smallest possible value if $\hat{b}_f(A) = b$. If the error function does not have an inherent preference for any partition (it's "homogeneous"), then the NFL theorem states [92, 96]

$$\sum_{(A,b)} \varepsilon(b, \hat{b}_f(A)) = \Lambda(\varepsilon), \tag{45}$$

where $\Lambda(\varepsilon)$ is a value that depends only on the error function chosen, but not on the community detection algorithm f. In other words, when averaged over all problem instances, all algorithms have the same accuracy. This implies, therefore, that in order for one class of algorithms to perform systematically better than another, we need to restrict the universe of problems to a particular subset. This is a seemingly straightforward result, but which is unfortunately very susceptible to misinterpretation and overstatement.

A common criticism of this kind of NFL theorem is that it is a poor representation of the typical problems we may encounter in real domains of application, which are unlikely to be uniformly distributed across the entire problem space.

Therefore, as soon as we constrain ourselves to a subset of problems that are relevant to a particular domain, then this will favor some algorithms over others — but then no algorithm will be superior for all domains. But since we are typically only interested in some domains, the NFL theorem is then arguably "theoretically sound, but practically irrelevant" [98]. Although indeed correct, in the case of community detection this logic is arguably an understatement. This is because as soon as we restrict our domain to community detection problems that reveal something *informative* about the network structure, then we are out of reach of the NFL theorem, and some algorithms will do better than others, without evoking any particular domain of application. We demonstrate this in the following.

The framework of the NFL theorem of Ref. [92] operates on a liberal notion of what constitutes a community detection problem and its solution, which means for an arbitrary pair (A, b) choosing the right f such that $\hat{b}_f(A) = b$. Under this framework, algorithms are just arbitrary mappings from network to partition, and there is no necessity to articulate more specifically how they relate to the structure of the network — community detection just becomes an arbitrary game of "guess the hidden node labels." This contrasts with how actual community detection algorithms are proposed, which attempt to match the node partitions to patterns in the network, e.g. assortativity, general connection preferences between groups, etc. Although the large variety of algorithms proposed for this task already reveal a lack of consensus on how to precisely define it, few would consider it meaningful to leave the class of community detection problems so wide open as to accept any matching between an arbitrary network and an arbitrary partition as a valid instance.

Even though we can accommodate any (deterministic) algorithm deemed valid according to any criterion under the NFL framework, most algorithms in this broader class do something else altogether. In fact, the absolute vast majority of them correspond to a maximally random matching between network and partition, which amounts to little more than just randomly guessing a partition for any given network, i.e. they return widely different partitions for inputs that are very similar, and overall point to no correlation between input and output.[22]

---

[22] An interesting exercise is to count how many such algorithms exist. A given community detection algorithm f needs to map each of all $\Omega(N) = 2^{\binom{N}{2}}$ networks of N nodes to one of $\Xi(N) = \sum_{B=1}^{N} {N \brace B} B!$ labeled partitions of its nodes. Therefore, if we restrict ourselves to a single value of N, the total number of input-output tables is $\Xi(N)^{\Omega(N)}$. If we sample one such table uniformly at random, it will be asymptotically impossible to compress it using fewer than $\Omega(N) \log_2 \Xi(N)$ bits — a number that grows super-exponentially with N. As an illustration, a random community detection algorithm that works only with $N = 100$ nodes would already need 10^{1479} terabytes of storage. Therefore, simply considering algorithms that humans can

It is not difficult to accept that these random algorithms perform equally "well" for any particular problem, or even all problems, but the NFL theorem says that they have equivalent performance even to algorithms that we may deem more meaningful. How do we make a formal distinction between algorithms that are just randomly guessing from those that are doing something coherent and trying to discover actual network patterns? As it turns out, there is an answer to this question that does not depend on particular domains of application: we require the solutions found to be *structured* and *compressive of the network.*

In order to interpret the statement of the NFL theorem in this vein, it is useful to re-write Eq. 45 using an equivalent probabilistic language,

$$\sum_{A,b} P(A,b)\varepsilon(b,\hat{b}_f(A)) = \Lambda'(\varepsilon), \tag{46}$$

where $\Lambda'(\varepsilon) \propto \Lambda(\varepsilon)$, and $P(A,b) \propto 1$ is the uniform probability of encountering a problem instance. When writing the theorem statement in this way, we notice immediately that instead of being agnostic about problem instances, it implies a *very specific* network generative model, which assumes a complete independence between network and partition. Namely, if we restrict ourselves to networks of N nodes, we have then:[23]

$$P(A,b) = P(A)P(b), \tag{47}$$

$$P(A) = 2^{-\binom{N}{2}}, \tag{48}$$

$$P(b) = \left[\sum_{B=1}^{N} \left\{{N \atop B}\right\} B! \right]^{-1}. \tag{49}$$

Therefore, the NFL theorem states simply that if we sample networks and partitions from a maximally random generative model, then all algorithms will have the same average accuracy at inferring the partition from the network. This is hardly a spectacular result — indeed the Bayes-optimal algorithm in this case, i.e. the one derived from the posterior distribution of the true generative model and which guarantees the best accuracy on average, consists of simply guessing partitions uniformly at random, ignoring the network structure altogether.

The probabilistic interpretation reveals that the NFL theorem involves a very specific assumption about what kind of community detection problem we are expecting. It is important to remember that it is not possible to make "no

---

write and use (together with their expected inputs and outputs) already pulls us very far away from the general scenario considered by the NFL theorem.

[23] We could easily introduce arbitrary constraints such as total number of edges or degree distribution, which would change the form of Eqs. 47 and 48, but none of the ensuing analysis.

assumption" about a problem; we are always forced to make *some* assumption, which even if implicit is not exempted from justification, and the uniform assumption of Eqs. 47 to 49 is no exception. In Fig. 14(a) we show a typical sample from this ensemble of community detection problems. In a very concrete sense, we can state that such problem instances are *unstructured* and contain *no learnable community structure*, or in fact no learnable network structure *at all*. We say that a community structure is (in principle) learnable if the knowledge of the partition b can be used to compress the network A, i.e. there exists an encoding $\mathcal{H}$ (i.e. a generative model) such that

$$\Sigma(A|b, \mathcal{H}) < -\log_2 P(A), \tag{50}$$

$$< \binom{N}{2}, \tag{51}$$

where $\Sigma(A|b, \mathcal{H}) = -\log_2 P(A|b, \mathcal{H})$ is the description length of A according to model $\mathcal{H}$, conditioned on the partition being known. However, it is a direct consequence of Shannon's source coding theorem [21, 29], that for the vast majority of networks sampled from the model of Eq. 47 the inequality in Eqs. (50) and (51) cannot be fulfilled as $N \to \infty$, i.e. the networks are incompressible.[24] This means that the true partition b carries no information about the network structure, and vice versa, i.e. the partition is not learnable from the network. In view of this, the common interpretation of the NFL theorem as "all algorithms perform equally well" is in fact quite misleading, and should be more accurately phrased as "all algorithms perform equally *poorly*," since no inferential algorithm can uncover the true community structure in most cases, at least no better than by chance alone. In other words, the universe of community detection problems considered in the NFL theorem is composed overwhelmingly of instances for which compression and explanation are not possible.[25] This uniformity between instances also reveals that there is no meaningful trade-off between algorithms for most instances, since all algorithms will yield

---

[24] For finite networks a positive compression might be achievable with small probability, but due to chance alone, and not in a manner that makes its structure learnable.

[25] One could argue that such a uniform model is justified by the principle of maximum entropy, which states that in the absence of prior knowledge about which problem instances are more likely, we should assume they are all equally likely *a priori*. This argument fails precisely because we *do* have sufficient prior knowledge that empirical networks are not maximally random — specially those possessing community structure, according to any meaningful definition of the term. Furthermore, it is easy to verify for each particular problem instance that the uniform assumption does not hold; either by compressing an observed network using any generative model (which should be asymptotically impossible under the uniform assumption [21]), or performing a statistical test designed to be able to reject the uniform null model. It is exceedingly difficult to find an empirical network for which the uniform model cannot be rejected with near-absolute confidence.

(a)   (b)

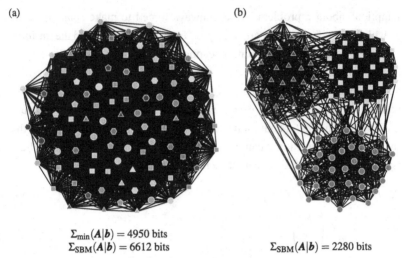

$$\Sigma_{\min}(\boldsymbol{A}|\boldsymbol{b}) = 4950 \text{ bits}$$
$$\Sigma_{\text{SBM}}(\boldsymbol{A}|\boldsymbol{b}) = 6612 \text{ bits}$$
$$\Sigma_{\text{SBM}}(\boldsymbol{A}|\boldsymbol{b}) = 2280 \text{ bits}$$

Figure 14 The NFL theorem involves predominantly instances of the community detection problem that are strictly incompressible, i.e. the true partitions cannot be used to explain the network. In (a) we show a typical sample of the uniform problem space given by Eq. 47, for $N = 100$ nodes, which yields a dense maximally random network, randomly divided into $B = 72$ groups. It is asymptotically impossible to use this partition to compress this network into fewer than $\Sigma_{\min}(\boldsymbol{A}|\boldsymbol{b}) = \binom{N}{2} = 4950$ bits, and therefore the partition is not learnable from the network alone with any inferential algorithm. We show also the description length of the SBM conditioned on the true partition, $\Sigma_{\text{SBM}}(\boldsymbol{A}|\boldsymbol{b})$, as a reference. In (b) we show an example of a community detection problem that is solvable, at least in principle, since $\Sigma_{\text{SBM}}(\boldsymbol{A}|\boldsymbol{b}) < \Sigma_{\min}(\boldsymbol{A}|\boldsymbol{b})$. In this case, the partition can be used to inform the network structure, and potentially vice-versa. This class of problem instance has a negligible contribution to the sum in the NFL theorem in Eq. 45, since it occurs only with an extremely small probability when sampled from the uniform model of Eq. 47. It is therefore more reasonable to state that the network in example (b) has an *actual* community structure, while the one in (a) does not.

the same negligible asymptotic performance, with an accuracy tending towards zero as the number of nodes increases. In this setting, there is not only no free lunch, but in fact there is no lunch at all (see Fig. 15).

If we were to restrict the space of possible community detection algorithms to those that provide actual explanations, then by definition this would imply a positive correlation between network and partition,[26] i.e.

---

[26] Note that Eq. 53 is a necessary but not sufficient condition for the community detection problem to be solvable. An example of this are networks generated by the SBM, which are solvable only if the strength of the community structure exceeds a detectability threshold [6], even if Eq. 53 is fulfilled.

$$P(A, b) = P(A|b)P(b) \tag{52}$$

$$\neq P(A)P(b). \tag{53}$$

Not only this implies a specific generative model but, as a consequence, also an *optimal* community detection algorithm, that operates based on the posterior distribution

$$P(b|A) = \frac{P(A|b)P(b)}{P(A)}. \tag{54}$$

Therefore, *learnable* community detection problems are invariably tied to an *optimal* class of algorithms, undermining to a substantial degree the relevance of the NFL theorem in practice. In other words, whenever there is an actual community structure in the network being considered — i.e. due to a systematic correlation between A and b, such that $P(A, b) \neq P(A)P(b)$ — there will be algorithms that can exploit this correlation better than others (see Fig. 14(b) for an example of a learnable community detection problem). Importantly, the set of learnable problems form only an infinitesimal fraction of all problem instances, with a measure that tends to zero as the number of nodes increases, and hence remain firmly out of scope of the NFL theorem. This observation has been made before, and is equally valid, in the wider context of NFL theorems beyond community detection [99–103].

Note that since there are many ways to choose a nonuniform model according to Eq. 53, the optimal algorithms will still depend on the particular assumptions made via the choice of $P(A, b)$ and how it relates to the true distribution. However, this does not imply that all algorithms have equal performance on compressible problem instances. If we sample a problem from the universe $\mathcal{H}_1$, with $P(A, b|\mathcal{H}_1)$, but use instead two algorithms optimal in $\mathcal{H}_2$ and $\mathcal{H}_3$, respectively, their relative performances will depend on how close each of these universes is to $\mathcal{H}_1$, and hence will not be in general the same. In fact, if our space of universes is finite, we can compose them into a single unified universe [104] according to

$$P(A, b) = \sum_{i=1}^{M} P(A, b|\mathcal{H}_i)P(\mathcal{H}_i), \tag{55}$$

which will incur a compression penalty of at most $\log_2 M$ bits added to the description length of the optimal algorithm. This gives us a path, based on hierarchical Bayesian models and minimum description length, to achieve optimal or near-optimal performance on instances of the community detection problem that are actually solvable, simply by progressively expanding our set of hypotheses.

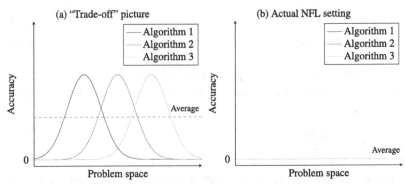

Figure 15 A common interpretation of the NFL theorem for community detection is that it reveals a necessary trade-off between algorithms: since they all have the same average performance, if one algorithm does better than another in one set of instances, it must do worse on a equal number of different instances, as depicted in panel (a). However, in the actual setting considered by the NFL theorem there is no meaningful trade-off: asymptotically, all algorithms perform maximally poorly for the vast majority of instances, as depicted in panel (b), since in these cases the network structure is uninformative of the partition. If we constrain ourselves to informative problem instances (which compose only an infinitesimal fraction of all instances), the NFL theorem is no longer applicable.

The idea that we can use compression as an inference criterion has been formalized by Solomonoff's theory of inductive inference [105], which forms a rigorous induction framework based on the principle of Occam's razor. Importantly, the expected errors of predictions achieved under this framework are provably upper-bounded by the Kolmogorov complexity of the data generating process [106], making the induction framework consistent. As we mentioned already in Sec. 2.3, the Kolmogorov complexity is a generalization of the description length we have been using, and it is defined by the length of the shortest binary program that generates the data. The only major limitation of Solomonoff's framework is its uncomputability, i.e. the impossibility of determining Kolmogorov's complexity with any algorithm [30]. However, this impossibility does not invalidate the framework, it only means that induction cannot be fully automated: we have a consistent criterion to compare hypotheses, but no deterministic mechanism to produce directly the best hypothesis. There are open philosophical questions regarding the universality of this inductive framework [107, 108], but whatever fundamental limitations it may have do not follow directly from NFL theorems such as the one from Ref. [92]. In fact, as mentioned in footnote 25, it is a rather simple task to use compression to reject the uniform hypothesis forming the basis of the NFL theorem for almost any network data.

Since compressive community detection problems are out of the scope of the NFL theorem, it is not meaningful to use it to justify avoiding comparisons between algorithms, on the grounds that all choices must be equally "good" in a fundamental sense. In fact, we do not need much sophistication to reject this line of argument, since the NFL theorem applies also when we are considering trivially inane algorithms, e.g. one that always returns the same partition for every network. The only domain where such an algorithm is as good as any other is when we have no community *structure* to begin with, which is precisely what the NFL theorem relies on.

Nevertheless, there are some lessons we can draw from the NFL theorem. It makes it clear that the performances of algorithms are tied directly to the inductive bias adopted, which should always be made explicit. The superficial interpretation of the NFL theorem as an inherent equity between all algorithms stems from the assumption that considering all problem instances uniformly is equivalent to being free of an inductive bias, but that is not possible. The uniform assumption is itself an inductive bias, and one that it is hard to justify in virtually any context, since it involves almost exclusively unsolvable problems (from the point of view of compressibility). In contrast, considering only *compressible* problem instances is also an inductive bias, but one that relies only on Occam's razor as a guiding principle. The advantage of the latter is that it is independent of domain of application, i.e. we are requiring only that an inferred partition can help explaining the network in some manner, without having to specify exactly how *a priori*.

In view of the above observations, it becomes easier to understand results such as of Ghasemian *et al* [66] who found that compressive inferential community detection methods tend to systematically outperform descriptive methods in empirical settings, when these are employed for the task of edge prediction. Even though edge prediction and community detection are not the same task, and using the former to evaluate the latter can lead in some cases to overfitting [109], typically the most compressive models will also lead to the best generalization. Therefore, the superior performance of the inferential methods is understandable, even though Ghasemian *et al* also found a minority of instances where some descriptive methods can outperform inferential ones. To the extent that these minority results cannot be attributed to overfitting, or technical issues such as insufficient MCMC equilibration, it could simply mean that the structure of these networks fall sufficiently outside of what is assumed by the inferential methods, but without it being a necessary trade-off that comes as a consequence of the NFL theorem — after all, under the uniform assumption, edge prediction is also strictly impossible, just like community detection. In other words, these results do not rule out the existence of an algorithm that works better in all cases considered, at least if their number is not too large.[27] In fact, this is precisely what is achieved in Ref. [110] via model stacking, i.e. a combination of several predictors into a meta-predictor that achieves systematically superior performance. This points indeed to the possibility of using universal methods to discover the latent *compressive* modular structure of networks, without any tension with the NFL theorem.

4.8 "Statistical inference requires us to believe the generative model being used."

We have been advocating for the use of statistical inference for community detection in networks whenever our objective is of an inferential nature.

One possible objection to the use of statistical inference is when the generative models on which they are based are considered unrealistic for a particular kind of network. Although this type of consideration is ultimately important, it is not necessarily an obstacle. First we need to remember that realism is a matter of degree, not kind, since no model can be fully realistic, and therefore we should never be fully committed to "believe" any particular model. Because of this, an inferential approach can be used to target a particular kind of structure, and the corresponding model is formulated with this in mind, but without the need to describe other properties of the data. The SBM is a good example of this, since it is often used with the objective of finding communities, rather than any kind of network structure. A model like the SBM is a good way to offset the regularities that relate to the community structure with the irregularities present in real networks, without requiring us to believe that in fact it generated the network.

Furthermore, certain kinds of models are flexible enough so that they can approximate other models. For example, a good analogy with fitting the SBM to network data is to fit a histogram to numerical data, with the node partitioning being analogous to the data binning. Although a piecewise constant

---

[27] It is important to distinguish the actual statement of the NFL theorem — "all algorithms perform equally well when averaged over all problem instances" — from the alternative statement: "No single algorithm exhibits strictly better performance than all others over all instances." Although the latter is a corollary of the former, it can also be true when the former is false. In other words, a particular algorithm can be better on average over relevant problem instances, but still underperform for some of them. In fact, it would only be possible for an algorithm to strictly dominate all others if it can always achieve perfect accuracy for every instance. Otherwise, there will be at least one algorithm (e.g. one that always returns the same partition) that can achieve perfect accuracy for a single network where the optimal algorithm does not ("even a broken clock is right twice a day"). Therefore, sub-optimal algorithms can eventually outperform optimal ones by chance when a sufficiently large number of instances is encountered, even when the NFL theorem is not applicable (and therefore this fact is not necessarily a direct consequence of it).

model is almost never the true underlying distribution, it provides a reasonable approximation in a tractable, nonparametric manner. Because of its capacity to approximate a wide class of distributions, we certainly do not need to believe that a histogram is the true data-generating process to extract meaningful inferences from it. In fact, the same can be said of the SBM in its capacity to approximate a wide class of network models [111, 112].

The above means that we can extract useful, statistically meaningful information from data even if the models we use are misspecified. For example, if a network is generated by a latent space model [113], and we fit a SBM to it, the communities that are obtained in this manner are not quite meaningless: they will correspond to discrete spatial regions. Hence, the inference would yield a caricature of the underlying latent space, amounting to a discretization of the true model — indeed, much like a histogram. This is very different from, say, finding communities in an Erdős-Rényi graph, which bear no relation to the true underlying model, and would be just overfitting the data. In contrast, the SBM fit to a spatial network would be approximately capturing the true model structure, in a manner that could be used to compress the data and make predictions (although not optimally).

Furthermore, the associated description length of a network model is a good criterion to tell whether the patterns we have found are actually simplifying our network description, without requiring the underlying model to be perfect. This happens in the same way as using a software like gzip makes our files smaller, without requiring us to believe that they are in fact generated by the Markov chain underlying the Lempel-Ziv algorithm [114].

Of course, realism becomes important as soon as we demand more from the point of view of interpretation and prediction. Are the observed community structures due to homophily or triadic clusure [23]? Or are they due to spatial embedding [113]? What models are capable of reproducing other network descriptors, together with the community structure? Which models can better reconstruct incomplete networks [53, 54]? When answering these questions, we are forced to consider more detailed generative processes, and compare them. However, we are never required to *believe* them — models are always tentative, and should always be replaced by superior alternatives when these are found. Indeed, criteria such as MDL serve precisely to implement such a comparison between models, following the principle of Occam's razor. Therefore, the lack of realism of any particular model cannot be used to dismiss statistical inference as an underlying methodology. On the contrary, the Bayesian workflow [115] enables a continuous improvement of our modelling apparatus, via iterative model building, model checking, and validation, all within a principled and consistent framework.

It should be emphasized that, fundamentally, there is no alternative. Rejecting an inferential approach based on the SBM on the grounds that it is an unrealistic model (e.g. because of the conditional independence of the edges being placed, or some other unpalatable assumption), but instead preferring some other non-inferential community detection method is incoherent: As we discussed in Sec. 2.5, every descriptive method can be mapped to an inferential analogue, with implicit assumptions that are hidden from view. Unless one can establish that the implicit assumptions are in fact more realistic, then the comparison cannot be justified. Unrealistic assumptions should be replaced by more realistic ones, not by burying one's head in the sand.

4.9 "Inferential approaches are prohibitively expensive."

One of the reasons why descriptive methods such as modularity maximization are widely used is because of very efficient heuristics that enable their application for very large networks. The most famous of which is the Louvain algorithm [116], touted for its speed and good ability to find high-scoring partitions. A more recent variation of this method is the Leiden algorithm [117], which is a refinement of the Louvain approach, designed to achieve even more high-scoring partitions, without sacrificing speed. None of these methods were developed with the purpose of assessing the statistical evidence of the partitions found, and since they are most often employed as modularity maximization techniques, they suffer from all the shortcomings that come with it.

It is often perceived that principled inferential approaches based on the SBM, designed to overcome all of the shortcomings of descriptive methods including modularity maximization, are comparatively much slower, often prohibitively so. However, we show here that this perception is quite inaccurate, since modern inferential approaches can be quite competitive. From the point of view of algorithmic complexity, agglomerative [118] or merge-split MCMC [119] have at most a log-linear complexity $O(E \log^2 N)$, where N and E are the number of nodes and edges, respectively, when employed to find the most likely partition. This means they belong to the same complexity class as the Louvain and Leiden algorithms, despite the fact the SBM-based algorithms are in fact more general, and do not attempt to find strictly assortative structures — and hence cannot make any optimizations that are only applicable in this case, as done by Louvain and Leiden. In practice, all these algorithms return results in comparable times.

In Fig. 16 we show a performance comparison between various algorithms on 38 empirical networks of various domains and number of edges spanning six orders of magnitude, obtained from the Netzschleuder repository [65]. We

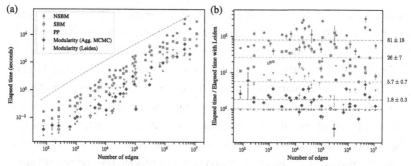

Figure 16 Inferential algorithms show competitive performance with descriptive ones. In panel (a) is shown the run-time of the Leiden algorithm [117] and the agglomerative MCMC [118] for modularity, and three SBM parametrizations: planted partition (PP), degree-corrected SBM, and nested degree-corrected SBM (NSBM), for 38 empirical networks [65]. All experiments were done on a laptop with an i9-9980HK Intel CPU, and averaged over at least 10 realizations. The dashed line shows an $O(E \log^2 E)$ scaling. In (b) are shown the same run times, but relative to the Leiden algorithm. The horizontal dashed lines show the median values.

used the Leiden implementation provided by its authors,[28] and compared with various SBM parametrizations implemented in the graph-tool library [9]. In particular we consider the agglomerative MCMC of Ref. [118] employed for modularity maximization, the Bayesian planted partition (PP) model [24], the degree-corrected SBM with uniform priors [16] and the nested SBM [16, 67]. As seen in Fig. 16(a), all algorithms display the same scaling with the number of edges, and differ only by an approximately constant factor. This difference is speed is due to the more complex likelihoods used by the SBM and additional data structures that are needed for its computation. When the agglomerative MCMC [118] is used with the simpler modularity function, it comes very close to the Leiden algorithm, despite not taking advantage of any custom optimization for that particular quality function. When used with the strictly assortative PP model, the algorithm slows down by a larger factor when compared to Leiden — most of which can be attributed to the increased complexity of the quality function. For the general SBM and nested SBM the algorithm slows down further, since now it is searching for arbitrary mixing patterns (not only assortative ones) and entire modular hierarchies. Indeed the performance difference between the most complex SBM and Leiden can be substantial, but at this point it also becomes an apples-and-oranges comparison, since the inferential method not only is not restricted to assortative communities, but it also

---

[28] Retreived from https://github.com/vtraag/leidenalg.

uncovers an entire hierarchy of partitions in a nonparametric manner, while being unhindered by the resolution limit and with protection against overfitting. Overall, if a practitioner is considering modularity maximization, they should prefer instead at least the Bayesian PP model, which solves the same kind of problem but it is not marred by all the shortcomings of modularity, including the resolution limit and systematic overfitting, while still being comparatively fast. The more advanced SBM formulations allow the researcher to probe a wider array of mixing patterns, without abdicating from statistical robustness, at the expense of increased computation time. As this analysis shows, all algorithms are accessible for fairly large networks of up to 10^7 edges on a laptop, but in fact can scale to 10^9 or more on high-performance computing (HPC) systems.

Based on the above, it becomes difficult to justify the use modularity maximization based solely on performance concerns, even on very large networks, since there are superior inferential approaches available with comparable speed, and which achieve more meaningful results in general.[29]

4.10 "Belief propagation outperforms MCMC."

The method of belief propagation (BP) [6] is an alternative algorithm to MCMC for inferring the partitions from the posterior distribution of the SBM in the semi-parametric case where the model parameters controlling the probability of connections between groups and the expected sizes of the groups are known *a priori*. It relies on the assumption that the network analyzed was truly sampled from the SBM, that the number of groups is much smaller than the number of nodes, $B \ll N$, and the network is sufficiently large, $N \gg 1$. Even though none of these assumptions are likely to hold in practice, BP is an extremely useful and powerful algorithm since it returns an estimate of the marginal posterior probability that is not stochastic, unlike MCMC. Furthermore, it is amenable to analytical investigations, which was used to uncover the detectability threshold of the SBM [6, 45], and important connections with spectral clustering [120]. It is often claimed, however, that it is also faster than MCMC when employed for the same task. This is, however, not quite

---

[29] In this comparison we consider only the task of finding point estimates, i.e. best scoring partitions. This is done to maintain an apples-to-apples comparison, since this all that can be obtained with the Leiden and other modularity maximization algorithms. To take full advantage of the Bayesian framework we would need to characterize the full posterior distribution instead, and sample partitions from it, instead of maximizing it, which incurs a larger computational cost and requires a more detailed analysis [26]. We emphasize, however, that the point estimates obtained with the SBM posterior already contain a substantial amount of regularization, and will not overfit the number of communities, for example.

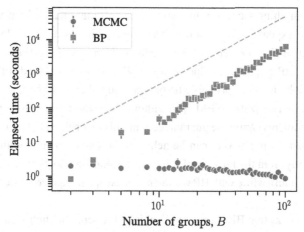

Figure 17 Comparison of run times between MCMC and BP on laptop with an i9-9980HK Intel CPU, for a network of flights between airports, with $N = 3188$ nodes and $E = 18833$. We used the agglomerative algorithm of Ref. [118], and initialized BP with the model parameters found with MCMC. The dashes line shows a B^2 slope.

true in general, as we now discuss. The complexity of BP is $O(\tau NB^2)$, where τ is the convergence time, which is typically small compared to the other quantities [for the DC-SBM the complexity becomes $O(\tau \ell NB^2)$, where ℓ is the number of distinct degrees in the network [27]]. A MCMC sweep of the SBM, i.e. the number of operations required to give a chance of each node to be moved once from its current node membership, can be implemented in time $O(N)$, independent of the number of groups B [118, 119], when using the parametrization of Refs. [16, 85]. This means that the performance difference between both approaches can be substantial when the number of groups is large. In fact, if $B = O(\sqrt{N})$ which a reasonable reference for empirical networks, BP becomes $O(N^2)$ while MCMC remains $O(N)$. Agglomerative MCMC initialization schemes, which can significantly improve the mixing time, have themselves a complexity $O(N \log^2 N)$ [118], still significantly faster than BP for large B.

In Fig. 17 we show a run-time comparison between BP and MCMC for an empirical network of flights between airports.[30] As the number of groups increases, the run-time of BP grows quadratically, as expected, while for MCMC it remains constant. There are several caveats in this comparison, which is somewhat apples-to-oranges: BP outputs a full marginal distribution for every node, containing even probabilities that are very low, while for MCMC

---

[30] Obtained from https://openflights.org/data.html.

we obtain anything from a point estimate to full marginal or joint probabilities, at the expense of longer running times, which is not revealed by the comparison in Fig. 17, which corresponds only to a point estimate. On the other hand, BP requires a value of the model parameters besides the partition itself, which can in principle be obtained together with the marginals via expectation-maximization (EM) [6], although a meaningful convergence for complex problems cannot be guaranteed with this algorithm [121]. Overall, we can state that some answers can be achieved in log-linear time with MCMC independently from the number of groups (and requiring no particular assumptions on the data), while with BP we can never escape the quadratic dependence on B.

We emphasize that BP is only applicable in the semiparametric case, where the number of groups and model parameters are known. The nonparametric case considered in Sec. 2.3, which is arguably more relevant in practice, cannot be tackled using BP, leaving MCMC as the only game in town, at least with the current state-of-the-art.

4.11 "Spectral clustering outperforms likelihood-based methods."

Spectral clustering methods divide a network into groups based on the leading eigenvectors of a linear operator associated with the network structure [122, 123]. There are important connections between spectral methods and statistical inference, in particular there are certain linear operators that can be shown to provide a consistent estimation of the SBM [120, 124]. However, when compared to likelihood-based methods, spectral methods are only approximations, as they amount to a simplification of the problem. Nevertheless, one of the touted advantages of this class of methods is that they tend to be significantly faster than likelihood based methods using MCMC. But like in the case of BP considered in the previous section, the run-time of spectral methods is intimately related to the number of groups one wishes to infer, unlike MCMC. Independently of the operator being used, the clustering into B groups requires the computation of the first B leading eigenvectors. The most efficient algorithms for this purpose are based on the implicitly restarted Arnoldi method [125], which has a worse-case time complexity $O(NB^2)$ for sparse matrices. Therefore, for sufficiently large number of groups they can cease to be faster than MCMC, which has a run-time complexity independent of the number of groups [118, 119].

In Fig. 18 we show a comparison of spectral clustering and MCMC inference for the Anybeat social network [126]. Indeed, for small number of groups

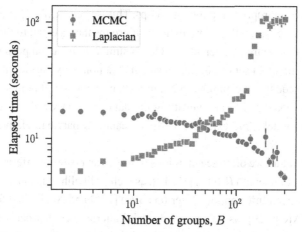

Figure 18 Comparison of run times between MCMC and spectral clustering using the Laplacian matrix, on a laptop with an i9-9980HK Intel CPU, for the Anybeat social network [126], with $N = 12645$ vertices and $E = 49132$ edges. We used the agglomerative algorithm of Ref. [118] and the ARPACK eigenvector solver [127].

spectral clustering can be significantly faster, but eventually becomes slower as the number of groups increases. The complexity of the spectral algorithm does not scale exactly like the worse case $O(NB^2)$ in practice, and the actual times will depend on the details of the particular operator. The MCMC algorithm becomes slightly faster, on the other hand, since the agglomerative initialization heuristic used terminates sooner when more groups are imposed [118]. As usual, there are caveats with this comparison. First, the eigenvectors by themselves do not provide a clustering of the network. Usually, these are given as input to a general-purpose clustering algorithm, typically k-means, which itself also has a complexity $O(NB^2)$, not included in the comparison of Fig. 18. Furthermore, spectral clustering usually requires the number of groups itself to be known in advance — although heuristics exist for spectral algorithms, but which usually require a significant part of the entire spectrum to be determined [120]. Likelihood-based methods, if implemented as a nonparametric Bayesian posterior like done in Sec. 2.3, do not require this prior information. On the other hand, spectral methods can be parallelized rather easily, unlike MCMC, and hence can take advantage of multicore processors.

4.12 "Bayesian posterior, MDL, BIC and AIC are different but equally valid model selection criteria."

One outstanding problem with using inferential community detection is that the likelihood of a model like the SBM does not, by itself, offer a principled way to determine the appropriate number of groups. This is because if we maximize the likelihood directly, it will favor a number of groups that is equal to the number of nodes, i.e. an extreme overfitting. This is similar to what happens when we fit a polynomial to a set of one-dimensional data points by varying its degree: for a degree equal to the number of points we can fit any set of points perfectly, but we are guaranteed to be overfitting the data. In other words, if we do not account for model complexity explicitly, we cannot separate randomness from structure.

In the literature we often see mentions of Bayesian posterior inference, minimum description length (MDL) [17, 18], as well as likelihood penalty schemes such as Bayesian Information Criterion (BIC) [128] and Akaike's Information Criterion (AIC)[129], as being equally valid alternatives that can be used to solve this problem. It is sometimes said that the choice between them is philosophical and often simply reflects the culture that a researcher stems from. As we show here, this is demonstrably incorrect, since Bayes, MDL, and BIC are in fact the same criterion, where BIC is simply an (arguably crude) approximation of the first two, which are in fact identical. AIC is indeed a different criterion, but, like BIC, it involves approximations that are known to be invalid for community detection.

The exact equivalence between MDL and Bayesian inference is easy to demonstrate [16, 67], as we have already done already in Sec. 2.3. Namely, the posterior distribution of the community detection problem is given by

$$P(b|A) = \frac{P(A|b)P(b)}{P(A)}, \tag{56}$$

$$= \frac{2^{-\Sigma(A,b)}}{P(A)}, \tag{57}$$

where the numerator of Eq. 56 is related to the description length $\Sigma(A, b)$ via

$$\Sigma(A, b) = -\log_2 P(A|b) - \log_2 P(b). \tag{58}$$

Therefore, maximizing Eq. 56 is identical to minimizing Eq. 58. Although this is already sufficient to demonstrate their equivalence, we can go in even more detail and show that the marginal integrated likelihood,

$$P(A|b) = \int P(A|\omega, \kappa, b)P(\omega, \kappa|b) \, d\omega \, d\kappa, \tag{59}$$

where ω and κ are the parameters of the canonical DC-SBM [15], is identical to the marginal likelihood of the microcanonical SBM we have used in Eq. 2. This is proved in Ref. [16]. Therefore, the MDL criterion is simply an information-theoretical interpretation of the Bayesian approach, and the two methods coincide in their implementation.[31]

The BIC criterion is based on the exact same framework, but it amounts to an approximation of the integrated marginal likelihood of a generic model $\mathcal{M}$, $P(D|\theta, \mathcal{M})$, where D is a data vector of size n and θ is a parameter vector of size k, given by

$$P(D|\mathcal{M}) = \int P(D|\theta, \mathcal{M})P(\theta)\,d\theta, \tag{60}$$

$$\approx \left(\frac{2\pi}{n}\right)^{k/2} |I(\hat{\theta})|\,\hat{L} \times P(\hat{\theta}), \tag{61}$$

$$\approx \exp(-\text{BIC}/2), \tag{62}$$

where $[I(\theta)]_{ij} = \int (\partial \ln P(D|\theta)/\partial\theta_i)(\partial \ln P(D|\theta)/\partial\theta_j)P(D|\theta)\,d\theta$ is the Fisher information matrix, and the values of the likelihood and parameters are obtained at the maximum,

$$\hat{L} = \max_{\theta} P(D|\theta, \mathcal{M}), \qquad \hat{\theta} = \underset{\theta}{\text{argmax}}\ P(D|\theta, \mathcal{M}), \tag{63}$$

and finally the BIC score is obtained from Eq. 61 by assuming $n \gg k$,

$$\text{BIC} = k \ln n - 2 \ln \hat{L}. \tag{64}$$

The BIC method consists of employing the equation above as criterion to decide which model to select, applicable even if they have different number of parameters k, with the first term functioning as penalty for larger models. Eq. 61 corresponds to an approximation of the likelihood obtained via Laplace's method, which involves a second-order Taylor expansion of the log-likelihood. Therefore, it requires the likelihood function to be well approximated by a multivariate Gaussian distribution with respect to the parameters at the vicinity of its maximum. However, as demonstrated by Yan *et al.* [27], this assumption is invalid for SBMs, however large the networks are, as long as they are *sparse*, i.e. with an average degree much smaller than the number of nodes. This is because for sparse SBMs we have both the number of parameters $k = O(N)$ [or even larger, since for B groups we a matrix ω of size $O(B^2)$, and in principle we could have $B = O(N)$] and effective data size $n = O(N)$ where N is the number of nodes, therefore the "sufficient data" limit required for the approximation to hold is never realized for any N. Furthermore, the BIC penalty completely neglects the contribution of the prior $P(\theta)$ in the regularization,

---

[31] In general, it is possible to construct particular MDL formulations of "universal codes" that do not have a clear Bayesian interpretation [18]. However, these formulations are typically intractable and seldom find an application. All MDL uses encountered in practice for the community detection problem are equivalent to Bayesian methods.

which cannot be ignored outside of this limit. Since the vast majority of empirical networks of interest are sparse, this renders this method unreliable, and in fact it will tend to overfit in most cases when employed with the SBM. We emphasize that the approximation of Eq. 61 is unnecessary, since we can compute the marginal likelihood of Eq. 59 exactly for most versions of the SBM [16, 54, 67, 130, 131]. When we compare the BIC penalty with the exact values of the integrated likelihoods we see that they in general produce significantly different regularizations, even asymptotically, and also even if we add *ad hoc* parameters, e.g. $\lambda k \ln n - 2 \ln \hat{L}$. This is because simply counting the number of parameters is too crude an estimation of the model complexity, since it is composed of different classes of parameters occupying different volumes which need (and can) be more carefully computed. Therefore the use of BIC for model selection in community detection should be in general avoided.

Akaike's Information Criterion (AIC) [129], on the other hand, actually starts out from a different framework. The idea is assume that the data are sampled from a true generative model $P(D|\mathcal{M}_{true})$, and a candidate model $\mathcal{M}$ with its parameter estimates $\hat{\theta}(D)$ is evaluated according to its Kullback-Leibler (KL) divergence with respect to the true model,

$$\int P(D'|\mathcal{M}_{true}) \ln \frac{P(D'|\hat{\theta}(D),\mathcal{M})}{P(D'|\mathcal{M}_{true})} \, dD'. \tag{65}$$

Of course, whenever it is relevant to employ model selection criteria we do not have access to the true model, which means we cannot compute the quantity in Eq. (65). We can, however, estimate the following upper bound, corresponding to the average over all data D,

$$\int P(D|\mathcal{M}_{true})P(D'|\mathcal{M}_{true}) \ln \frac{P(D'|\hat{\theta}(D),\mathcal{M})}{P(D'|\mathcal{M}_{true})} \, dD' \, dD. \tag{66}$$

In this case, for sufficiently large data D, the quantity in Eq. (66) can be estimated making use of a series of Laplace approximations [132], resulting in

$$\ln P(D|\hat{\theta}(D)) - \mathrm{tr}\left[J(\theta_0)I(\theta_0)^{-1}\right], \tag{67}$$

where θ_0 is the point around which we compute the quadratic approximation in Laplace's method, and $J_{ij}(\theta_0) = \int P(D'|\mathcal{M}_{true})\mathcal{I}_{ij}(D,\theta_0) \, dD$, $I_{ij}(\theta_0) = \int P(D'|\theta_0,\mathcal{M})\mathcal{I}_{ij}(D,\theta_0) \, dD$, with

$$\mathcal{I}_{ij}(D,\hat{\theta}) = \left.\frac{\partial}{\partial\theta_i} \ln P(D'|\theta,\mathcal{M})\right|_{\theta_i=\hat{\theta}_i} \times \left.\frac{\partial}{\partial\theta_j} \ln P(D'|\theta,\mathcal{M})\right|_{\theta_j=\hat{\theta}_j}. \tag{68}$$

The AIC criterion is finally obtained by heuristically assuming $\mathrm{tr}\left[J(\boldsymbol{\theta}_0)I(\boldsymbol{\theta}_0)^{-1}\right] \approx k$, yielding

$$\text{AIC} = 2k - 2\ln P(\boldsymbol{D}|\hat{\boldsymbol{\theta}}(\boldsymbol{D})), \tag{69}$$

where the overall sign and multiplicative factor is a matter of convention. It is also possible to recover AIC from BIC by making a choice of prior $P(\mathcal{M}) \propto \exp(k \ln n/2 - k)$ [132], which makes it clear that it favors more complex models over BIC. Independently of how one judges the suitability of the fundamental criterion of Eq. 66, just like BIC, AIC involves several approximations that are known to be invalid for sparse networks. Together with its heuristic nature and crude counting of parameters, it is safe to conclude that the use of AIC is ill-advised for community detection, specially considering the more principled and exact alternatives of Bayes/MDL.

5 Conclusion

We have framed the problem of community detection under two different paradigms, namely that of "inference" and "description." We argued that statistical inference is unavoidable when the objective is to draw inferential interpretations from the communities found, and we provided a simple "litmus test" to help deciding when this is indeed the case. Under this framing, we showed that descriptive methods always come with hidden inferential assumptions, and reviewed the dangers of employing descriptive methods with inferential aims, focusing on modularity maximization as a representative (and hence not unique) case.

We covered a series of pitfalls encountered in community detection, as well as myths and half-truths commonly believed, and attempted to clarify them under the same lenses, focusing on simple examples and conceptual arguments.

Although it is true that community detection in general involves diverse aims, and hence it is difficult to argue for an one-size-fits-all approach, here we have taken a more opinionated stance, since it is also not true that all approaches are used in a manner consistent with their intended aims. We have clearly favored inferential methods, since they are more theoretically grounded, are better aligned with well-defined scientific questions (whenever those involve inferential queries), are more widely applicable, and can be used to develop more robust algorithms.

Inferential methodology for community detection has reached a level of maturity, both in our understanding of them and in the efficiency of available implementations, that should make it the preferred choice when analysing network data, whenever the ultimate goal has an inferential nature.

References

[1] Santo Fortunato, "Community detection in graphs," Physics Reports **486**, 75–174 (2010).

[2] Santo Fortunato and Darko Hric, "Community detection in networks: A user guide," Physics Reports (2016).

[3] Cristopher Moore, "The Computer Science and Physics of Community Detection: Landscapes, Phase Transitions, and Hardness," arXiv:1702.00467 (2017).

[4] Emmanuel Abbe, "Community detection and stochastic block models: recent developments," arXiv:1703.10146 [cs, math, stat] (2017).

[5] Tiago P. Peixoto, "Bayesian Stochastic Blockmodeling," in *Advances in Network Clustering and Blockmodeling*, edited by P. Doreian, V. Batagelj, and A. Ferligoj (John Wiley & Sons, Ltd, 2019) pp. 289–332.

[6] Aurelien Decelle, Florent Krzakala, Cristopher Moore, and Lenka Zdeborová, "Asymptotic analysis of the stochastic block model for modular networks and its algorithmic applications," Physical Review E **84**, 066106 (2011).

[7] Lenka Zdeborová and Florent Krzakala, "Statistical physics of inference: thresholds and algorithms," Advances in Physics **65**, 453–552 (2016).

[8] Michael T. Schaub, Jean-Charles Delvenne, Martin Rosvall, and Renaud Lambiotte, "The many facets of community detection in complex networks," Applied Network Science **2**, 1–13 (2017).

[9] Tiago P. Peixoto, "The graph-tool python library," figshare (2014), 10.6084/m9.figshare.1164194, available at `https://graph-tool.skewed.de`.

[10] R. Jacob Baker, *CMOS: Circuit Design, Layout, and Simulation*, 3rd ed. (Wiley-IEEE Press, Piscataway, NJ : Hoboken, NJ, 2010).

[11] Brian Wilson Kernighan, *Some graph partitioning problems related to program segmentation* (Princeton University, 1969).

[12] B.W. Kernighan and S. Lin, "An efficient heuristic procedure for partitioning graphs," Bell System Technical Journal **49**, 291–307 (1970).

[13] Charles-Edmond Bichot and Patrick Siarry, *Graph partitioning* (John Wiley & Sons, 2013).

[14] Paul W. Holland, Kathryn Blackmond Laskey, and Samuel Leinhardt, "Stochastic blockmodels: First steps," Social Networks **5**, 109–137 (1983).

[15] Brian Karrer and M. E. J. Newman, "Stochastic blockmodels and community structure in networks," Physical Review E **83**, 016107 (2011).

[16] Tiago P. Peixoto, "Nonparametric Bayesian inference of the microcanonical stochastic block model," Physical Review E **95**, 012317 (2017).

[17] J. Rissanen, "Modeling by shortest data description," Automatica **14**, 465–471 (1978).

[18] Peter D. Grünwald, *The Minimum Description Length Principle* (The MIT Press, 2007).

[19] Jorma Rissanen, *Information and Complexity in Statistical Modeling*, 1st ed. (Springer, 2010).

[20] David J. C. MacKay, *Information Theory, Inference and Learning Algorithms*, first edition ed. (Cambridge University Press, 2003).

[21] C. E Shannon, "A mathematical theory of communication," Bell Syst Tech. J **27**, 623 (1948).

[22] Marinka Zitnik, Rok Sosič, Marcus W. Feldman, and Jure Leskovec, "Evolution of resilience in protein interactomes across the tree of life," Proceedings of the National Academy of Sciences **116**, 4426–4433 (2019), publisher: National Academy of Sciences Section: PNAS Plus.

[23] Tiago P. Peixoto, "Disentangling Homophily, Community Structure, and Triadic Closure in Networks," Physical Review X **12**, 011004 (2022).

[24] Lizhi Zhang and Tiago P. Peixoto, "Statistical inference of assortative community structures," Physical Review Research **2**, 043271 (2020).

[25] Romualdo Pastor-Satorras, Eric Smith, and Ricard V. Solé, "Evolving protein interaction networks through gene duplication," Journal of Theoretical Biology **222**, 199–210 (2003).

[26] Tiago P. Peixoto, "Revealing Consensus and Dissensus between Network Partitions," Physical Review X **11**, 021003 (2021).

[27] Xiaoran Yan, Cosma Shalizi, Jacob E. Jensen, Florent Krzakala, Cristopher Moore, Lenka Zdeborová, Pan Zhang, and Yaojia Zhu, "Model selection for degree-corrected block models," Journal of Statistical Mechanics: Theory and Experiment **2014**, P05007 (2014).

[28] Aurelien Decelle, Florent Krzakala, Cristopher Moore, and Lenka Zdeborová, "Phase transition in the detection of modules in sparse networks," 1102.1182 (2011).

[29] Thomas M. Cover and Joy A. Thomas, *Elements of Information Theory*, 99th ed. (Wiley-Interscience, 1991).

[30] Ming Li and Paul M. B. Vitányi, *An Introduction to Kolmogorov Complexity and Its Applications*, 3rd ed. (Springer, New York, 2008).

[31] Tom A. B. Snijders and Krzysztof Nowicki, "Estimation and Prediction for Stochastic Blockmodels for Graphs with Latent Block Structure," Journal of Classification **14**, 75–100 (1997).

[32] Krzysztof Nowicki and Tom A. B Snijders, "Estimation and Prediction for Stochastic Blockstructures," Journal of the American Statistical Association **96**, 1077–1087 (2001).

[33] Christian Tallberg, "A Bayesian Approach to Modeling Stochastic Blockstructures with Covariates," The Journal of Mathematical Sociology **29**, 1–23 (2004).

[34] M. B. Hastings, "Community detection as an inference problem," Physical Review E **74**, 035102 (2006).

[35] Martin Rosvall and Carl T. Bergstrom, "An information-theoretic framework for resolving community structure in complex networks," Proceedings of the National Academy of Sciences **104**, 7327–7331 (2007).

[36] Edoardo M. Airoldi, David M. Blei, Stephen E. Fienberg, and Eric P. Xing, "Mixed Membership Stochastic Blockmodels," J. Mach. Learn. Res. **9**, 1981–2014 (2008).

[37] Aaron Clauset, Cristopher Moore, and M. E. J. Newman, "Hierarchical structure and the prediction of missing links in networks," Nature **453**, 98–101 (2008).

[38] Jake M. Hofman and Chris H. Wiggins, "Bayesian Approach to Network Modularity," Physical Review Letters **100**, 258701 (2008).

[39] Morten Mørup and Lars Kai Hansen, "Learning latent structure in complex networks," in *NIPS Workshop on Analyzing Networks and Learning with Graphs* (2009).

[40] Marián Boguñá and Romualdo Pastor-Satorras, "Class of correlated random networks with hidden variables," Physical Review E **68**, 036112 (2003).

[41] Béla Bollobás, Svante Janson, and Oliver Riordan, "The phase transition in inhomogeneous random graphs," Random Structures & Algorithms **31**, 3–122 (2007).

[42] Andrea Lancichinetti, Santo Fortunato, and Filippo Radicchi, "Benchmark graphs for testing community detection algorithms," Physical Review E **78**, 046110 (2008).

[43] M. Girvan and M. E. J. Newman, "Community structure in social and biological networks," Proceedings of the National Academy of Sciences **99**, 7821 –7826 (2002).

[44] Andrea Lancichinetti and Santo Fortunato, "Community detection algorithms: A comparative analysis," Physical Review E **80**, 056117 (2009).

[45] Aurelien Decelle, Florent Krzakala, Cristopher Moore, and Lenka Zdeborová, "Inference and Phase Transitions in the Detection of Modules in Sparse Networks," Physical Review Letters **107**, 065701 (2011).

[46] M. E. J. Newman, "Modularity and community structure in networks," Proceedings of the National Academy of Sciences **103**, 8577–8582 (2006).

[47] Martin Rosvall and Carl T. Bergstrom, "Maps of random walks on complex networks reveal community structure," Proceedings of the National Academy of Sciences **105**, 1118–1123 (2008).

[48] R. Lambiotte, J. C. Delvenne, and M. Barahona, "Random Walks, Markov Processes and the Multiscale Modular Organization of Complex Networks," IEEE Transactions on Network Science and Engineering **1**, 76–90 (2014).

[49] Andrew Gelman, John B. Carlin, Hal S. Stern, David B. Dunson, Aki Vehtari, and Donald B. Rubin, *Bayesian Data Analysis*, 3rd ed. (Chapman and Hall/CRC, Boca Raton, 2013).

[50] Christopher M. Bishop, *Pattern Recognition and Machine Learning* (Springer, 2011).

[51] M. E. J. Newman, "Network structure from rich but noisy data," Nature Physics **14**, 542–545 (2018).

[52] Travis Martin, Brian Ball, and M. E. J. Newman, "Structural inference for uncertain networks," Physical Review E **93**, 012306 (2016).

[53] Tiago P. Peixoto, "Reconstructing Networks with Unknown and Heterogeneous Errors," Physical Review X **8**, 041011 (2018).

[54] Roger Guimerà and Marta Sales-Pardo, "Missing and spurious interactions and the reconstruction of complex networks," Proceedings of the National Academy of Sciences **106**, 22073 –22078 (2009).

[55] Till Hoffmann, Leto Peel, Renaud Lambiotte, and Nick S. Jones, "Community detection in networks without observing edges," Science Advances **6**, eaav1478 (2020), publisher: American Association for the Advancement of Science Section: Research Article.

[56] Tiago P. Peixoto, "Network Reconstruction and Community Detection from Dynamics," Physical Review Letters **123**, 128301 (2019).

[57] B. Fosdick, D. Larremore, J. Nishimura, and J. Ugander, "Configuring Random Graph Models with Fixed Degree Sequences," SIAM Review **60**, 315–355 (2018).

[58] Fan Chung and Linyuan Lu, "Connected Components in Random Graphs with Given Expected Degree Sequences," Annals of Combinatorics **6**, 125–145 (2002).

[59] Roger Guimerà, Marta Sales-Pardo, and Luís A. Nunes Amaral, "Modularity from fluctuations in random graphs and complex networks," Physical Review E **70**, 025101 (2004).

[60] Santo Fortunato and Marc Barthélemy, "Resolution limit in community detection," Proceedings of the National Academy of Sciences **104**, 36–41 (2007).

[61] Benjamin H. Good, Yves-Alexandre de Montjoye, and Aaron Clauset, "Performance of modularity maximization in practical contexts," Physical Review E **81**, 046106 (2010).

[62] M. E. J. Newman, "Mixing patterns in networks," Phys. Rev. E **67**, 026126 (2003).

[63] Maria A. Riolo and M. E. J. Newman, "Consistency of community structure in complex networks," Physical Review E **101**, 052306 (2020).

[64] Lizhi Zhang and T. P. Peixoto, "Large-scale assessment of overfitting, underfitting and model selection for modular network structures," in preparation.

[65] T. P. Peixoto, "The Netzschleuder network catalogue and repository." (2020), accessible at https://networks.skewed.de.

[66] Amir Ghasemian, Homa Hosseinmardi, and Aaron Clauset, "Evaluating Overfit and Underfit in Models of Network Community Structure," IEEE Transactions on Knowledge and Data Engineering , 1–1 (2019).

[67] Tiago P. Peixoto, "Hierarchical Block Structures and High-Resolution Model Selection in Large Networks," Physical Review X **4**, 011047 (2014).

[68] Daniel B. Larremore, Aaron Clauset, and Abigail Z. Jacobs, "Efficiently inferring community structure in bipartite networks," Physical Review E **90**, 012805 (2014).

[69] Xiao Zhang, Travis Martin, and M. E. J. Newman, "Identification of core-periphery structure in networks," Physical Review E **91**, 032803 (2015).

[70] Pan Zhang and Cristopher Moore, "Scalable detection of statistically significant communities and hierarchies, using message passing for modularity," Proceedings of the National Academy of Sciences **111**,

18144–18149 (2014), publisher: National Academy of Sciences Section: Physical Sciences.

[71] M. E. J. Newman, "Equivalence between modularity optimization and maximum likelihood methods for community detection," Physical Review E **94** (2016), 10.1103/PhysRevE.94.052315.

[72] Jörg Reichardt and Stefan Bornholdt, "Statistical mechanics of community detection," Physical Review E **74**, 016110 (2006).

[73] A. Arenas, A. Fernández, and S. Gómez, "Analysis of the structure of complex networks at different resolution levels," New Journal of Physics **10**, 053039 (2008).

[74] Peter J. Bickel and Aiyou Chen, "A nonparametric view of network models and Newman–Girvan and other modularities," Proceedings of the National Academy of Sciences **106**, 21068–21073 (2009).

[75] M. E. J. Newman, "Spectral methods for community detection and graph partitioning," Physical Review E **88**, 042822 (2013).

[76] Claire P. Massen and Jonathan P. K. Doye, "Thermodynamics of Community Structure," arXiv:cond-mat/0610077 (2006).

[77] Andrea Lancichinetti and Santo Fortunato, "Consensus clustering in complex networks," Scientific Reports **2**, 1–7 (2012), number: 1 Publisher: Nature Publishing Group.

[78] Jörg Reichardt and Stefan Bornholdt, "When are networks truly modular?" Physica D: Nonlinear Phenomena **224**, 20–26 (2006).

[79] Dandan Hu, Peter Ronhovde, and Zohar Nussinov, "Phase transitions in random Potts systems and the community detection problem: spin-glass type and dynamic perspectives," Philosophical Magazine **92**, 406–445 (2012).

[80] Alec Kirkley and M. E. J. Newman, "Representative community divisions of networks," Communications Physics **5**, 1–10 (2022), number: 1 Publisher: Nature Publishing Group.

[81] David V. Foster, Jacob G. Foster, Peter Grassberger, and Maya Paczuski, "Clustering drives assortativity and community structure in ensembles of networks," Physical Review E **84**, 066117 (2011).

[82] Andrea Lancichinetti and Santo Fortunato, "Limits of modularity maximization in community detection," Physical Review E **84**, 066122 (2011).

[83] Clara Granell, Sergio Gómez, and Alex Arenas, "Hierarchical multiresolution method to overcome the resolution limit in complex networks," International Journal of Bifurcation and Chaos **22**, 1250171 (2012).

[84] Tatsuro Kawamoto and Martin Rosvall, "Estimating the resolution limit of the map equation in community detection," Physical Review E **91**, 012809 (2015).

[85] Tiago P. Peixoto, "Parsimonious Module Inference in Large Networks," Physical Review Letters **110**, 148701 (2013).

[86] Michael J Barber, "Modularity and community detection in bipartite networks," 0707.1616 (2007).

[87] Mel MacMahon and Diego Garlaschelli, "Community Detection for Correlation Matrices," Physical Review X **5**, 021006 (2015).

[88] V. A. Traag and Jeroen Bruggeman, "Community detection in networks with positive and negative links," Physical Review E **80**, 036115 (2009).

[89] Paul Expert, Tim S. Evans, Vincent D. Blondel, and Renaud Lambiotte, "Uncovering space-independent communities in spatial networks," Proceedings of the National Academy of Sciences **108**, 7663–7668 (2011), publisher: National Academy of Sciences Section: Physical Sciences.

[90] Darko Hric, Tiago P. Peixoto, and Santo Fortunato, "Network Structure, Metadata, and the Prediction of Missing Nodes and Annotations," Physical Review X **6**, 031038 (2016).

[91] M. E. J. Newman and Aaron Clauset, "Structure and inference in annotated networks," Nature Communications **7**, 11863 (2016).

[92] Leto Peel, Daniel B. Larremore, and Aaron Clauset, "The ground truth about metadata and community detection in networks," Science Advances **3**, e1602548 (2017).

[93] Y. Hu, "Efficient, high-quality force-directed graph drawing," Mathematica Journal **10**, 37–71 (2005).

[94] Andreas Noack, "Modularity clustering is force-directed layout," Physical Review E **79**, 026102 (2009).

[95] David H. Wolpert and William G. Macready, *No free lunch theorems for search*, Tech. Rep. (Technical Report SFI-TR-95-02-010, Santa Fe Institute, 1995).

[96] David H. Wolpert, "The Lack of A Priori Distinctions Between Learning Algorithms," Neural Computation **8**, 1341–1390 (1996).

[97] David H. Wolpert and William G. Macready, "No free lunch theorems for optimization," IEEE transactions on evolutionary computation **1**, 67–82 (1997).

[98] Cullen Schaffer, "A Conservation Law for Generalization Performance," in *Machine Learning Proceedings 1994*, edited by William W. Cohen and Haym Hirsh (Morgan Kaufmann, San Francisco (CA), 1994) pp. 259–265.

[99] Matthew J. Streeter, "Two Broad Classes of Functions for Which a No Free Lunch Result Does Not Hold," in *Genetic and Evolutionary Computation — GECCO 2003*, Lecture Notes in Computer Science, edited by Erick Cantú-Paz, James A. Foster, Kalyanmoy Deb, Lawrence David Davis, Rajkumar Roy, Una-May O'Reilly, Hans-Georg Beyer, Russell Standish, Graham Kendall, Stewart Wilson, Mark Harman, Joachim Wegener, Dipankar Dasgupta, Mitch A. Potter, Alan C. Schultz, Kathryn A. Dowsland, Natasha Jonoska, and Julian Miller (Springer, Berlin, Heidelberg, 2003) pp. 1418–1430.

[100] Simon McGregor, "No free lunch and algorithmic randomness," in *GECCO*, Vol. 6 (2006) pp. 2–4.

[101] Tom Everitt, "Universal induction and optimisation: No free lunch?" unpublished master's thesis, Stockholms Universitet (2013).

[102] Tor Lattimore and Marcus Hutter, "No Free Lunch versus Occam's Razor in Supervised Learning," in *Algorithmic Probability and Friends. Bayesian Prediction and Artificial Intelligence: Papers from the Ray Solomonoff 85th Memorial Conference, Melbourne, VIC, Australia, November 30 – December 2, 2011*, Lecture Notes in Computer Science, edited by David L. Dowe (Springer, Berlin, Heidelberg, 2013) pp. 223–235.

[103] Gerhard Schurz, *Hume's Problem Solved: The Optimality of Meta-Induction*, illustrated edition ed. (The MIT Press, Cambridge, Massachusetts, 2019).

[104] E. T. Jaynes, *Probability Theory: The Logic of Science*, edited by G. Larry Bretthorst (Cambridge University Press, Cambridge, UK; New York, NY, 2003).

[105] R. J. Solomonoff, "A formal theory of inductive inference. Part I," Information and Control 7, 1–22 (1964).

[106] Marcus Hutter, "On universal prediction and Bayesian confirmation," Theoretical Computer Science Theory and Applications of Models of Computation, **384**, 33–48 (2007).

[107] Marcus Hutter, "Open Problems in Universal Induction & Intelligence," Algorithms **2**, 879–906 (2009), number: 3 Publisher: Molecular Diversity Preservation International.

[108] George D. Montanez, "Why machine learning works," unpublished Ph.D. thesis, Carnegie Mellon University, Pittsburgh (2017).

[109] Toni Vallès-Català, Tiago P. Peixoto, Marta Sales-Pardo, and Roger Guimerà, "Consistencies and inconsistencies between model selection and link prediction in networks," Physical Review E **97**, 062316 (2018).

[110] Amir Ghasemian, Homa Hosseinmardi, Aram Galstyan, Edoardo M. Airoldi, and Aaron Clauset, "Stacking models for nearly optimal link prediction in complex networks," Proceedings of the National Academy of Sciences **117**, 23393–23400 (2020).

[111] Sofia C. Olhede and Patrick J. Wolfe, "Network histograms and universality of blockmodel approximation," Proceedings of the National Academy of Sciences **111**, 14722–14727 (2014).

[112] Jean-Gabriel Young, Guillaume St-Onge, Patrick Desrosiers, and Louis J. Dubé, "Universality of the stochastic block model," Physical Review E **98**, 032309 (2018).

[113] Peter D Hoff, Adrian E Raftery, and Mark S Handcock, "Latent Space Approaches to Social Network Analysis," Journal of the American Statistical Association **97**, 1090–1098 (2002).

[114] J. Ziv and A. Lempel, "A universal algorithm for sequential data compression," IEEE Transactions on Information Theory **23**, 337–343 (1977).

[115] Andrew Gelman, Aki Vehtari, Daniel Simpson, Charles C. Margossian, Bob Carpenter, Yuling Yao, Lauren Kennedy, Jonah Gabry, Paul-Christian Bürkner, and Martin Modrák, "Bayesian Workflow," (2020), arXiv:2011.01808.

[116] Vincent D. Blondel, Jean-Loup Guillaume, Renaud Lambiotte, and Etienne Lefebvre, "Fast unfolding of communities in large networks," Journal of Statistical Mechanics: Theory and Experiment **2008**, P10008 (2008).

[117] V. A. Traag, L. Waltman, and N. J. van Eck, "From Louvain to Leiden: guaranteeing well-connected communities," Scientific Reports **9**, 5233 (2019).

[118] Tiago P. Peixoto, "Efficient Monte Carlo and greedy heuristic for the inference of stochastic block models," Physical Review E **89**, 012804 (2014).

[119] Tiago P. Peixoto, "Merge-split Markov chain Monte Carlo for community detection," Physical Review E **102**, 012305 (2020).

[120] Florent Krzakala, Cristopher Moore, Elchanan Mossel, Joe Neeman, Allan Sly, Lenka Zdeborová, and Pan Zhang, "Spectral redemption in clustering sparse networks," Proceedings of the National Academy of Sciences, **110**, 20935–20940 (2013).

[121] Tatsuro Kawamoto, "Algorithmic detectability threshold of the stochastic block model," Physical Review E **97**, 032301 (2018).

[122] Daniel A. Spielman and Shang-Hua Teng, "Spectral partitioning works: planar graphs and finite element meshes," Linear Algebra and its Applications **421**, 284–305 (2007).

[123] Ulrike von Luxburg, "A tutorial on spectral clustering," Statistics and Computing **17**, 395–416 (2007).

[124] Karl Rohe, "Spectral clustering and the high-dimensional stochastic blockmodel," The Annals of Statistics **39**, 1878–1915 (2011).

[125] R. B. Lehoucq and D. C. Sorensen, "Deflation Techniques for an Implicitly Restarted Arnoldi Iteration," SIAM Journal on Matrix Analysis and Applications **17**, 789–821 (1996).

[126] Michael Fire, Rami Puzis, and Yuval Elovici, "Link Prediction in Highly Fractional Data Sets," in *Handbook of Computational Approaches to Counterterrorism*, edited by V.S. Subrahmanian (Springer, New York, NY, 2013) pp. 283–300.

[127] Richard B. Lehoucq, Danny C. Sorensen, and Chao Yang, *ARPACK users' guide: solution of large-scale eigenvalue problems with implicitly restarted Arnoldi methods* (SIAM, 1998).

[128] Gideon Schwarz, "Estimating the Dimension of a Model," The Annals of Statistics **6**, 461–464 (1978).

[129] H. Akaike, "A new look at the statistical model identification," IEEE Transactions on Automatic Control **19**, 716–723 (1974).

[130] Etienne Côme and Pierre Latouche, "Model selection and clustering in stochastic block models based on the exact integrated complete data likelihood," Statistical Modelling **15**, 564–589 (2015).

[131] M. E. J. Newman and Gesine Reinert, "Estimating the Number of Communities in a Network," Physical Review Letters **117**, 078301 (2016).

[132] Kenneth P. Burnham and David R. Anderson, eds., *Model Selection and Multimodel Inference: A Practical Information-Theoretic Approach* (Springer, New York, NY, 2002).

Cambridge Elements ☰

The Structure and Dynamics of Complex Networks

Guido Caldarelli
Ca' Foscari University of Venice

Guido Caldarelli is Full Professor of Theoretical Physics at Ca' Foscari University of Venice. Guido Caldarelli received his Ph.D. from SISSA, after which he held postdoctoral positions in the Department of Physics and School of Biology, University of Manchester, and the Theory of Condensed Matter Group, University of Cambridge. He also spent some time at the University of Fribourg in Switzerland, at École Normale Supérieure in Paris, and at the University of Barcelona. His main scientific activity (interest?) is the study of networks, mostly analysis and modelling, with applications from financial networks to social systems as in the case of disinformation. He is the author of more than 200 journal publications on the subject, and three books, and is the current President of the Complex Systems Society (2018 to 2021).

About the Series

This cutting-edge series provides authoritative and detailed coverage of the underlying theory of complex networks, specifically their structure and dynamical properties. Each Element within the series will focus upon one of three primary topics: static networks, dynamical networks, and numerical/computing network resources.

Cambridge Elements ≡

The Structure and Dynamics of Complex Networks

Elements in the Series

Reconstructing Networks
Giulio Cimini, Rossana Mastrandrea and Tiziano Squartini

Higher-Order Networks
Ginestra Bianconi

The Shortest Path to Network Geometry: A Practical Guide to Basic Models and Applications
M. Ángeles Serrano and Marián Boguñá

Weak Multiplex Percolation
Gareth J. Baxter, Rui A. da Costa, Sergey N. Dorogovtsev and José F. F. Mendes

Modularity and Dynamics on Complex Networks
Renaud Lambiotte and Michael Schaub

Percolation in Spatial Networks
Bnaya Gross and Shlomo Havlin

Multilayer Network Science
Oriol Artime, Barbara Benigni, Giulia Bertagnolli, Valeria d'Andrea, Riccardo Gallotti, Arsham Ghavasieh, Sebastian Raimondo and Manlio De Domenico

Gillespie Algorithms for Stochastic Multiagent Dynamics in Populations and Networks
Naoki Masuda and Christian L. Vestergaard

Descriptive vs. Inferential Community Detection in Networks
Tiago P. Peixoto

A full series listing is available at: www.cambridge.org/SDCN

Printed in the United States
by Baker & Taylor Publisher Services